Navigation and Tracking in Space

Analysis and Algorithms

For a complete listing of titles in the
Artech House *GNSS Technology and Applications* series,
turn to the back of this book.

Navigation and Tracking in Space

Analysis and Algorithms

Sanat K. Biswas

Andrew G. Dempster

ARTECH HOUSE

BOSTON | LONDON
artechhouse.com

Library of Congress Cataloging-in-Publication Data
A catalog record for this book is available from the U.S. Library of Congress.

British Library Cataloguing in Publication Data
A catalogue record for this book is available from the British Library.

Cover design by Joi Garron

ISBN 13: 978-1-63081-920-0

685 Canton Street
Norwood, MA 02062

10 9 8 7 6 5 4 3 2 1

Contents

Preface

This book is the culmination of a decade-long journey in the field of space vehicle navigation, bringing together our experiences and research findings. Our goal is to present the concepts and methods used in space vehicle navigation and tracking in a concise manner, specifically targeting young engineers and scientists who aspire to venture into this fascinating domain.

In recent years, the world has witnessed a renewed and invigorated interest in space exploration, driven by a combination of technological advancements, international collaborations, and the quest for scientific discovery and human expansion beyond Earth's boundaries. Governments, private companies, and research institutions are investing in ambitious missions to explore distant planets, asteroids, and celestial bodies.

The space navigation and tracking landscape has changed significantly as a result. With the launch of multiple mega-constellations and frequent interplanetary missions, faster, a more accurate, and more frequent navigation and tracking solution is now a requirement. One objective of this book is to address this requirement. The authors are recognized experts in space vehicle navigation and Global Navigation Satellite Systems (GNSS). The computationally efficient estimators included within these pages, namely the single propagation unscented Kalman filter (SPUKF), extrapolated single propagation unscented Kalman filter (ESPUKF), and extrapolated single propagation particle filter (ESP-PF), are the outcome of the authors' collaborative research.

The central part of this book adopts a tutorial approach, providing clear explanations accompanied by example codes. The accompanied codes are essentially end-to-end system and observation simulation as well as implementation of various estimation algorithms for navigation and tracking of launch, reentry, Earth-orbiting satellites, as well as interplanetary spacecraft. We believe that this approach will empower our readers to construct their own navigation and tracking algorithms from scratch and build upon existing codes. By providing a solid foundation, we aim to inspire innovation and foster the development of novel navigation and tracking solutions.

Our focus throughout the book lies in the navigation and tracking of artificial space objects. We explore a broad spectrum of space missions, encompassing earth-orbiting satellite missions, launch and reentry missions, as well as interplanetary missions. Each mission type necessitates the modeling of dynamics and the design of estimation algorithms tailored to its specific requirements.

We delve into the intricacies of force models, accounting for perturbations caused by nonuniform gravitational fields, tidal effects, atmospheric drag, solar radiation pressure, and gravitational influences from other celestial bodies. Furthermore, we discuss various reference frames utilized for navigation in space. To facilitate the implementation of navigation filters, we provide state space models for satellites, launch vehicles, reentry vehicles, and interplanetary spacecraft.

For earth-orbiting satellite tracking and navigation, we explore diverse ground and space-based observations, accompanied by detailed error models. Furthermore, we outline methodologies for achieving precise orbit determination. To illustrate the practical application of these concepts, we present navigation and tracking results using the extended Kalman filter (EKF) with simulated observations. We also address the numerical challenges that arise during the implementation of navigation algorithms.

The dynamics of launch and reentry vehicles exhibit extreme nonlinearity, necessitating a distinct treatment in navigation and tracking. Within this context, we include our recent research on the effects of nonlinearity on navigation performance and describe methods for mitigating estimation errors within the framework of the Kalman filter. Additionally, we introduce navigation and tracking techniques employing the unscented Kalman filter, particle filter, and their computationally efficient variants. Moreover, we discuss the selection of estimation algorithms based on the nonlinearity of the dynamics and considerations for real-time onboard navigation.

In crafting this book, we drew inspiration from the works of esteemed experts in the field, such as David Vallado, Oliver Montenbruck, B. D. Tapley, Yaakov Bar-Shalom, and many others. We owe much of our own understanding of astrodynamics and orbit determination to their invaluable contributions. Within these pages, we have consciously endeavored to blend the concepts of estimation theory with the context of astrodynamics, focusing specifically on navigation and tracking.

Our hope is that this book will serve as a valuable resource, empowering readers to:

- Design estimation algorithms tailored to space missions.

- Address nonlinearity in the design of navigation algorithms for space missions.
- Select suitable estimators for specific mission requirements.
- Develop space vehicle navigation algorithms for resource-constrained environments.

We trust that the information and insights presented within these chapters will spark innovation, inspire further research, and enable the next generation of engineers and scientists to embark on their own remarkable journeys in the realm of space navigation and tracking.

Chapter 1

Introduction

1.1 BACKGROUND

The launching of Sputnik marked the dawn of the space age in 1957. Apart from achieving scientific goals and technology demonstration, the Sputnik mission created scientific strides that resulted in lunar missions, earth observation satellites, the global navigation satellite system (GNSS), and numerous interplanetary missions. With time, the fleet of satellites deployed by various nations has become essential for day-to-day life. Numerous space missions are conceived and executed every year for navigation, communication, remote sensing, weather forecasting, resource management, astronomical observation, and solar system exploration. With these vital applications, space technology has proven to be the essential infrastructure of human civilization. It is anticipated that future space missions will open endless possibilities for technological advancements and shape the society of the coming age.

The operation of a space vehicle relies on a collection of interdependent functions: structure, power, navigation, attitude determination, guidance, control, thermal, telemetry and command, propulsion, and data handling. Among these vital functional segments, accurate navigation (i.e. determination of the position and velocity) is often crucial for the success of a space mission.

Precise position and velocity information of a launch vehicle is required for the insertion of a spacecraft into its orbit and for range safety. In-orbit position determination is important for station keeping, guidance, and manoeuvring of earth-orbiting artificial satellites and deep space-faring spacecraft. For reentry missions, position and velocity information are even more vital for proper reentry procedures and vehicle recovery.

Similar to navigation, tracking also involves the estimation of the position and velocity of spacecraft. Although the mathematical framework is similar for both navigation and tracking, they are often distinguished by where the observations and decisions are being made. In navigation, the observations are made by the onboard sensors, and the position and velocity estimation are also performed in the onboard flight computer. On the other hand, in tracking the observations are made from the ground station. The position and velocity estimation is done on the ground station as well in the case of tracking. Essentially a space vehicle, some measurement mechanism, and a processing unit to perform the estimation comprise a navigation or tracking system.

Navigation and tracking of all types of space vehicles require some input observations that are functions of the position and velocity of space vehicles. Dead-reckoning, range, range rate, pointing angles, and angular measurements to known celestial objects are widely used as measurements for space vehicle navigation and tracking. The position and velocity of space vehicles are estimated from these measurements using a suitable estimator.

Estimation techniques can be broadly segregated into two different categories: batch and sequential estimators. Batch estimators process observations from multiple epochs at once and provide navigation solutions. On the other hand, sequential estimators process observations epoch-by-epoch; that provides an updated navigation solution when a new set of observations is recorded. Sequential estimators require only the current set of observations to be stored, hence memory requirement is less. In addition, sequential estimators incorporate a priori information of the state vector to be estimated and provide better accuracy. As a result, these are preferred for real-time navigation and tracking.

Of several types of sequential estimators, the Kalman filter is designed for the state estimation of stochastic dynamic systems. It is a very popular estimation technique due to its computational efficiency and it is frequently used in space vehicle navigation and attitude estimation. The extended Kalman filter (EKF) was developed to apply the Kalman filter framework in nonlinear systems [1]. The application of the EKF spans almost all engineering disciplines. However, this algorithm provides suboptimal estimation for mildly nonlinear problems [2, 3] due to the first-order Taylor series approximation of the mean and conditional error covariance [4]. It has long been established that the degree of nonlinearity of a dynamic system is one of the decisive factors for the accuracy of the EKF. To address nonlinearity, several techniques involving analytical and numerical computation of the Jacobian and Hessian were developed [3, 5]. Julier et al. suggested a deterministic sampling approach to compute the a priori mean state vector and the error covariance to capture the nonlinearity of the dynamic system [6, 7]. This approach is known as the unscented Kalman filter and is a popular estimation technique for so-called highly nonlinear dynamic systems. Another notable sequential estimator is the particle filter where the probability density function (PDF) is represented as a set of random samples [8]. A large number of samples ensures an equivalent representation of the PDF. The random sample state vectors are propagated over the observation interval to obtain the a priori information. Once observations are received, the weights of each propagated sample state vector are computed using the likelihood function and a posteriori PDF of the state vector is determined using the weights. The particle filter is more accurate in nonlinear propagation of the PDF; however, it is also computationally very expensive.

The cost of launching a spacecraft is high and depends on the type of mission. For example, the cost of sending an artificial satellite into an LEO is around $2000 per kg. Hence, it is always desirable to minimize the mass of the space vehicle, which results in limited availability of computational resources onboard. Hence computationally expensive estimation algorithms often cannot be used for onboard navigation. A designer of a space vehicle navigation system must consider this major design constraint while choosing an estimation algorithm. An estimation algorithm with high accuracy but fewer numerical operations is the prudent choice.

The EKF is utilized in precise position estimation by post-processing on the ground [9–11] as well as for real-time onboard satellite position estimation [9, 12, 13] and formation flying of satellites using GNSS measurements [14]. Possible application of the UKF has been explored in satellite navigation, attitude determination and control [15, 16], GPS/INS integration for unmanned aerial vehicles [17], indoor positioning [18], target tracking [19], and in various other estimation problems. Although the UKF and PF provide a more accurate estimation solution than the EKF, these estimators are often not the preferred technique for real-time estimation due to their computationally expensive nature. We have provided the application of recently developed computationally efficient variants of the UKF and PF in space vehicle navigation and tracking in the forthcoming chapters.

The general qualitative notion of so-called highly nonlinear and mildly nonlinear systems is not sufficient for designers to forecast the accuracy improvement by the UKF over the EKF for a given set of system and measurement models. In [20] it was demonstrated that for a reentry vehicle tracking problem using range measurements, the UKF performance is superior to the EKF. In [21] the UKF was applied in the low earth orbit (LEO) satellite navigation using Global Positioning System (GPS) measurements. Despite being a nonlinear system with nonlinear measurements, it was observed that the performance improvement by the UKF compared to the EKF was not significant. Based on [22] we have discussed the selection of the suitable estimator based on some quantitative notion of nonlinearity in Chapter 4.

1.2 SCOPE OF THE BOOK

1.2.1 Objectives

The foremost objective of the book is to present the fundamental concepts, methods, and tools that are used in the navigation and tracking of various space vehicles and aid the readers to design and synthesis estimators for various space missions.

With this objective in mind, we have presented the mathematical framework of navigation and tracking problems in the first few chapters. We have presented the mathematical model of various space vehicles and observation models in the subsequent chapters.

The book also presents applications of various estimation algorithms in satellite navigation and tracking, launch vehicle and reentry vehicle tracking, and tracking of interplanetary spacecraft. The book also provides the application of recently developed computationally efficient variants of the conventional nonlinear estimators in space vehicle navigation and tracking.

The example applications are constructed under certain assumptions and simplifications to aid the readers to understand the application of the estimation algorithms for diverse types of space missions. We hope that with the basic building blocks and tutorial presented in the book, the readers will be easily able to explore the application of the presented algorithms in more realistic scenarios.

We will pose the navigation and tracking problem as a stochastic estimation problem which requires some preliminary background in linear algebra, vector calculus, stochastic process, and probability.

Section 1.3.1 provides a brief overview of linear algebra and state space method which provides the mathematical structure for posing estimation problems. Section 1.3.2 introduces the readers to the rotation matrices that are used to transform vectors from one reference frame to another. Section 1.3.3 reviews the basic concepts of probability theory and stochastic processes that are required to understand the stochastic estimation problem.

We assume the readers of the book are already familiar with these concepts. However, for completeness, we present very selected concepts and definitions briefly in this chapter. Should the readers require a deeper understanding of these topics, they can refer to books by Pierre Bremaud for probability theory and stochastic processes, and Norman Nise for state space method.

Additionally, we will discuss Kepler's Laws and orbital elements in Section 1.3.4 as preliminary before discussing the space vehicle dynamics in Chapter 2.

1.2.2 Overview of the Book

The rest of this chapter covers the preliminaries that are required to construct estimation problems. Chapter 2 provides a detailed process of developing the dynamic model of various space vehicles. The chapter starts by introducing the readers to various reference systems and frames which are key components in defining the position and velocity of space vehicles. Subsequently, the force model and perturbations are introduced. Finally, the state space models of a satellite, launch vehicle, reentry vehicle, and interplanetary spacecraft are provided. The prerequisite for this chapter is Chapter 1.

Chapter 3 presents various observation models that are used in the navigation and tracking of space vehicles. The chapter covers observation models for various ground-based observations, onboard observations GNSS and INS. Chapter 2 is a prerequisite for this chapter.

Chapter 4 presents various estimation algorithms that are used for navigation and tracking. The chapter starts with the basic concepts of estimation theory and then presents the Kalman filter and its variants. The chapter also presents the particle filter and its variants. The chapter concludes by describing the impact of nonlinearity in the estimator performance. Chapter 1 is the prerequisite for this chapter.

Chapter 5 provides a step-by-step procedure to implement sequential estimators for the navigation and tracking of a satellite. Two example applications are provided. The first example is a geostationary satellite tracking problem using ground-based observations. The second example is a LEO satellite navigation problem using GNSS measurements. The second example is further extended to examine the estimator performance with a change in the orbit eccentricity. Chapters 1 through 4 are the prerequisites for this chapter.

Chapter 6 provides a sequential estimator implementation strategy for reentry vehicle and launch vehicle tracking problems. An example of ground-based tracking of reentry vehicles is provided in this chapter. Further, the GPS-based launch vehicle navigation problem is presented in this chapter. Chapters 1 through 4 are the prerequisites for this chapter.

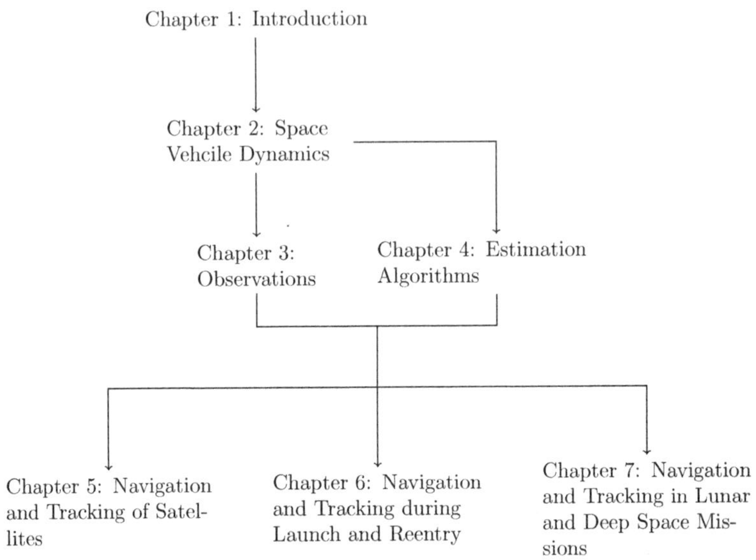

Figure 1.1: Book structure.

Chapter 7 presents the sequential estimator implementation strategy for tracking interplanetary spacecraft. The chapter starts with a description of the dynamics of interplanetary spacecraft. The chapter then presents the sequential estimator implementation strategy for tracking of interplanetary spacecraft using ground-based observations. Chapters 1 through 4 are the prerequisites for this chapter.

The book structure indicating the prerequisites of each chapter is shown in Figure 1.1.

1.3 PRELIMINARIES

1.3.1 Linear Algebra

Linear algebra forms the foundation of many mathematical and computational techniques used in various fields, including navigation and tracking in space. In this section, we will cover the key concepts and techniques of linear algebra that are essential for a thorough understanding of advanced topics in space navigation.

This section serves as a refresher for readers who are already familiar with linear algebra. Readers who are not familiar with linear algebra are encouraged to refer to the book by Gilbert Strang [23] for a detailed understanding of the subject.

1.3.1.1 Vector Space and Vector

A vector space $\mathbb{V}$ is a collection of objects called vectors that satisfy certain properties. These vectors can be thought of as elements or points in the vector space. The vector space is defined over a field, which is typically the real numbers ($\mathbb{R}$) or complex numbers ($\mathbb{C}$). In this book, we denote a vector using bold font.

Formally, a vector space is a set $\mathbb{V}$ together with two operations: vector addition and scalar multiplication. The vector addition operation takes two vectors from $\mathbb{V}$ and produces a new vector in $\mathbb{V}$, while scalar multiplication takes a scalar from the field and a vector from $\mathbb{V}$ and produces a new vector in $\mathbb{V}$.

For a vector space, the following additional properties must hold:

1. Associativity of addition: The sum of three vectors is independent of the grouping ; that is, $(\boldsymbol{u}+\boldsymbol{v})+\boldsymbol{w}=\boldsymbol{u}+(\boldsymbol{v}+\boldsymbol{w})$ for all $\boldsymbol{u},\ \boldsymbol{v},\ \boldsymbol{w} \in \mathbb{V}$.

2. Commutativity of addition: The sum of two vectors is independent of the order ; that is, $\boldsymbol{u}+\boldsymbol{v}=\boldsymbol{v}+\boldsymbol{u}$ for all $\boldsymbol{u},\ \boldsymbol{v} \in \mathbb{V}$.

3. Existence of zero vector: There exists a special vector called the zero vector, denoted as $\mathbf{0}$, which acts as an additive identity ; that is, $\boldsymbol{v}+\mathbf{0}=\boldsymbol{v}$ for all $\boldsymbol{v} \in \mathbb{V}$.

4. Existence of additive inverse: For every vector $\boldsymbol{v} \in \mathbb{V}$, there exists a vector $-\boldsymbol{v} \in \mathbb{V}$ such that $\boldsymbol{v}+(-\boldsymbol{v})=\mathbf{0}$.

5. Existence of multiplicative identity: There exists a special scalar called the multiplicative identity, denoted as 1, which acts as a multiplicative identity ; that is, $1\boldsymbol{v}=\boldsymbol{v}$ for all $\boldsymbol{v} \in \mathbb{V}$.

6. Distributivity of scalar multiplication over vector addition: For any scalar c and any vectors $\boldsymbol{u}, \boldsymbol{v} \in \mathbb{V}$, $c(\boldsymbol{u}+\boldsymbol{v})=c\boldsymbol{u}+c\boldsymbol{v}$.

7. Distributivity of scalar multiplication over field addition: For any scalars c, d and any vector $\boldsymbol{v} \in \mathbb{V}$, $(c+d)\boldsymbol{v}=c\boldsymbol{v}+d\boldsymbol{v}$.

8. Compatibility of scalar multiplication with field multiplication: For any scalars c, d and any vector $\boldsymbol{v} \in \mathbb{V}$, $(cd)\boldsymbol{v}=c(d\boldsymbol{v})$.

1.3.1.2 Linear Independence

A set of vectors is said to be linearly independent if none of the vectors in the set can be expressed as a linear combination of the others. In other words, a set of vectors $v_1, v_2, ..., v_n$ is linearly independent if the only solution to the equation:

$$c_1 v_1 + c_2 v_2 + ... + c_n v_n = 0 \tag{1.1}$$

is the trivial solution; that is, all the scalars $c_1, c_2, ..., c_n$ are zero ($c_1 = c_2 = ... = c_n = 0$). If a set of vectors is not linearly independent, it is called linearly dependent. In this case, at least one vector in the set can be expressed as a linear combination of the others.

1.3.1.3 Matrix and Matrix Operations

A matrix is an array of numbers arranged in rows and columns. Formally, an $m \times n$ matrix $\boldsymbol{A}$ is defined as a collection of mn elements arranged in m rows and n columns. The elements of a matrix can be real numbers, complex numbers, or other mathematical entities. The element in the i-th row and j-th column is denoted as a_{ij} or $A[i, j]$, where i ranges from 1 to m and j ranges from 1 to n. For example, a 2 matrix can be represented as:

$$\boldsymbol{A} = \begin{bmatrix} a_{11} & a_{12} & a_{13} \\ a_{21} & a_{22} & a_{23} \end{bmatrix} \tag{1.2}$$

here, $\boldsymbol{A}$ has 2 rows and 3 columns, and $a_{11}, a_{12}, a_{13}, a_{21}, a_{22}$, and a_{23} are the elements of the matrix. It should be noted that a set of matrices with fixed size or dimension forms (denoted as $\mathbb{R}^{m \times n}$) is a vector space, which means all the properties of vector space are applicable to matrices.

For an $m \times n$ matrix $\boldsymbol{A}$, then the *transpose* of $\boldsymbol{A}$, denoted as $\boldsymbol{A}^T$, is an $n \times m$ matrix obtained by interchanging the rows and columns of A. The element in the i-th row and j-th column of the transpose is the element in the j-th row and i-th column of the original matrix; that is, $A^T[j, i] = A[i, j]$. For the previously defined matrix $\boldsymbol{A}$ the transpose $\boldsymbol{A}^T$; that is:

$$\boldsymbol{A}^T = \begin{bmatrix} a_{11} & a_{21} \\ a_{12} & a_{22} \\ a_{13} & a_{23} \end{bmatrix} \tag{1.3}$$

Given two matrices $\boldsymbol{A} \in \mathbb{R}^{m \times n}$ and $\boldsymbol{B} \in \mathbb{R}^{n \times p}$, the matrix product $\boldsymbol{AB}$, denoted as $\boldsymbol{C} = \boldsymbol{AB}$, where $\boldsymbol{C} \in \mathbb{R}^{m \times p}$ and the elements of $\boldsymbol{C}$ are given by:

$$c_{ij} = \sum_{k=1}^{n} a_{ik} b_{kj} \tag{1.4}$$

If $\boldsymbol{A}$ is a square matrix ($\in \mathbb{R}^{n \times n}$) and there exists another matrix $\boldsymbol{A}^{-1}$ such that $\boldsymbol{A}\boldsymbol{A}^{-1} = \boldsymbol{A}^{-1}\boldsymbol{A} = \boldsymbol{I}_{n \times n}$, then $\boldsymbol{A}^{-1}$ is the inverse of $\boldsymbol{A}$. Here $\boldsymbol{I}_{n \times n}$ is the identity matrix of size $n \times n$ and is defined as:

$$\boldsymbol{I}_{n \times n} = \begin{bmatrix} 1 & 0 & \cdots & 0 \\ 0 & 1 & \cdots & 0 \\ \vdots & \vdots & \ddots & \vdots \\ 0 & 0 & \cdots & 1 \end{bmatrix} \tag{1.5}$$

The trace of a square matrix $\boldsymbol{A} \in \mathbb{R}^{n \times n}$, denoted as $tr(\boldsymbol{A})$, is the sum of the diagonal elements; that is

$$tr(\boldsymbol{A}) = \sum_{i=1}^{n} a_{ii} \tag{1.6}$$

The trace of a matrix has some important properties:

1. The trace is a linear operator, meaning that for matrices $\boldsymbol{A}$ and $\boldsymbol{B}$ and a scalar c, $tr(\boldsymbol{A} + \boldsymbol{B}) = tr(\boldsymbol{A}) + tr(\boldsymbol{B})$ and $tr(c\boldsymbol{A}) = c * tr(\boldsymbol{A})$.
2. The trace is invariant under cyclic permutations, which means that $tr(\boldsymbol{ABC}) = tr(\boldsymbol{CAB}) = tr(\boldsymbol{BCA})$, where $\boldsymbol{A}$, $\boldsymbol{B}$, and $\boldsymbol{C}$ are compatible matrices.

The rank of a matrix $\boldsymbol{A}$, denoted as $rank(\boldsymbol{A})$, is defined as the maximum number of linearly independent rows or columns in the matrix. It can also be interpreted as the maximum number of nonzero rows or columns that can be extracted from the matrix.

Given a square matrix $\boldsymbol{A}$, an eigenvector is a nonzero vector $\boldsymbol{v}$ such that when $\boldsymbol{A}$ is multiplied by $\boldsymbol{v}$, the resulting vector is parallel to $\boldsymbol{v}$, although it may be scaled by a scalar value. This scalar value is called the eigenvalue corresponding to that eigenvector.

Mathematically, we can represent this relationship as:

$$\boldsymbol{A}\boldsymbol{v} = \lambda \boldsymbol{v} \tag{1.7}$$

Here, λ is the corresponding eigenvalue. To find the eigenvalues and eigenvectors, we solve the characteristic equation:

$$|\boldsymbol{A} - \lambda \boldsymbol{I}| = 0 \tag{1.8}$$

where $\boldsymbol{I}$ is the identity matrix of the same size as $\boldsymbol{A}$. The eigenvalues are the roots of the characteristic equation. The eigenvectors are the vectors that satisfy the equation $\boldsymbol{Av} = \lambda \boldsymbol{v}$ for each eigenvalue. It can be shown that the trace of a matrix is equal to the sum of its eigenvalues. Another important property of eigenvalues is that the determinant of a matrix is equal to the product of its eigenvalues.

1.3.1.4 Gradient, Jacobian, and Hessian

The *gradient* of a scaler valued function $f : \mathbb{R}^n \to \mathbb{R}$ with respect to a vector $\boldsymbol{x} \in \mathbb{R}^n$ is defined as

$$\nabla_{\boldsymbol{x}} f(\boldsymbol{x}) = \begin{bmatrix} \frac{\partial f(\boldsymbol{x})}{\partial x_1} \\ \frac{\partial f(\boldsymbol{x})}{\partial x_2} \\ \vdots \\ \frac{\partial f(\boldsymbol{x})}{\partial x_n} \end{bmatrix} \tag{1.9}$$

The *gradient* of a vector valued function $\boldsymbol{f} : \mathbb{R}^n \to \mathbb{R}^m$ with respect to a vector $\boldsymbol{x} \in \mathbb{R}^n$ is defined as

$$\nabla_{\boldsymbol{x}} \boldsymbol{f}(\boldsymbol{x})^T = \begin{bmatrix} \frac{\partial f_1(\boldsymbol{x})}{\partial x_1} & \frac{\partial f_2(\boldsymbol{x})}{\partial x_1} & \cdots & \frac{\partial f_m(\boldsymbol{x})}{\partial x_1} \\ \vdots & \vdots & \ddots & \vdots \\ \frac{\partial f_1(\boldsymbol{x})}{\partial x_n} & \frac{\partial f_2(\boldsymbol{x})}{\partial x_n} & \cdots & \frac{\partial f_m(\boldsymbol{x})}{\partial x_n} \end{bmatrix} \tag{1.10}$$

The *Jacobian* of $\boldsymbol{f}$ is defined as the transpose of the gradient of $\boldsymbol{f}$. That means the *Jacobian* of $\boldsymbol{f}$ is a $m \times n$ matrix:

$$\frac{\partial \boldsymbol{f}(\boldsymbol{x})}{\partial \boldsymbol{x}} = \begin{bmatrix} \frac{\partial f_1(\boldsymbol{x})}{\partial x_1} & \frac{\partial f_1(\boldsymbol{x})}{\partial x_2} & \cdots & \frac{\partial f_1(\boldsymbol{x})}{\partial x_n} \\ \vdots & \vdots & \ddots & \vdots \\ \frac{\partial f_m(\boldsymbol{x})}{\partial x_1} & \frac{\partial f_m(\boldsymbol{x})}{\partial x_2} & \cdots & \frac{\partial f_m(\boldsymbol{x})}{\partial x_n} \end{bmatrix} \tag{1.11}$$

The *Hessian* of a scalar-valued function $f : \mathbb{R}^n \to \mathbb{R}$ with respect to a vector $\boldsymbol{x} \in \mathbb{R}^n$ is defined as

$$\frac{\partial^2 f(\boldsymbol{x})}{\partial \boldsymbol{x}^2} = \begin{bmatrix} \frac{\partial^2 f(\boldsymbol{x})}{\partial x_1^2} & \frac{\partial^2 f(\boldsymbol{x})}{\partial x_1 \partial x_2} & \cdots & \frac{\partial^2 f(\boldsymbol{x})}{\partial x_1 \partial x_n} \\ \frac{\partial^2 f(\boldsymbol{x})}{\partial x_2 \partial x_1} & \frac{\partial^2 f(\boldsymbol{x})}{\partial x_2^2} & \cdots & \frac{\partial^2 f(\boldsymbol{x})}{\partial x_2 \partial x_n} \\ \vdots & \vdots & \ddots & \vdots \\ \frac{\partial^2 f(\boldsymbol{x})}{\partial x_n \partial x_1} & \frac{\partial^2 f(\boldsymbol{x})}{\partial x_n \partial x_2} & \cdots & \frac{\partial^2 f(\boldsymbol{x})}{\partial x_n^2} \end{bmatrix} \tag{1.12}$$

The textitHessian for a vector-valued function is a tensor of order 3.

1.3.1.5 State Space Representation

Mechanical systems in general can be modeled using Newton's law of motion or the Euler-Lagrange equation, which eventually results in a differential equation. Depending on the system in question, the order of the differential equation may vary. State space representation of a system is nothing but reformatting a n-th order differential equation that represents the system to n first-order differential equations and represents using a single first-order vector differential equation. For example, consider the following linear differential equation:

$$\dddot{y} + \alpha\ddot{y} + \beta\dot{y} + \gamma y = 0 \tag{1.13}$$

where $y \in \mathbb{R}$. We can consider the state vector as $[y\ \dot{y}\ \ddot{y}]^T$. Then the state space representation of the above differential equation becomes

$$\begin{bmatrix} \dot{y} \\ \ddot{y} \\ \dddot{y} \end{bmatrix} = \begin{bmatrix} 0 & 1 & 0 \\ 0 & 0 & 1 \\ -\gamma & -\beta & -\alpha \end{bmatrix} \begin{bmatrix} y \\ \dot{y} \\ \ddot{y} \end{bmatrix} \tag{1.14}$$

which is essentially a first-order differential equation. This can be extended to nonlinear differential equations as well. We will use this state space representation for modelling the space vehicle motion.

1.3.1.6 State Transition Matrix

Consider a state space representation of a system as

$$\dot{\boldsymbol{X}}(t) = \boldsymbol{f}(\boldsymbol{X}, t) \tag{1.15}$$

The state transition matrix $\boldsymbol{\Phi}$ is defined as

$$\boldsymbol{\Phi}(t) = \frac{\partial \boldsymbol{X}(t)}{\partial \boldsymbol{X}(t_0)} \tag{1.16}$$

From (1.15) one can write

$$\frac{\partial}{\partial \boldsymbol{X}(t_0)}\dot{\boldsymbol{X}}(t) = \frac{\partial \boldsymbol{f}(\boldsymbol{X},t)}{\partial \boldsymbol{X}(t_0)} \tag{1.17}$$

which implies

$$\frac{d}{dt}\frac{\partial \boldsymbol{X}(t)}{\partial \boldsymbol{X}(t_0)} = \frac{\partial \boldsymbol{f}(\boldsymbol{X},t)}{\partial \boldsymbol{X}(t)}\frac{\partial \boldsymbol{X}(t)}{\partial \boldsymbol{X}(t_0)} \tag{1.18}$$

Then from the definition of the state transition matrix

$$\dot{\boldsymbol{\Phi}}(t) = \frac{\partial \boldsymbol{f}(\boldsymbol{X},t)}{\partial \boldsymbol{X}(t)}\boldsymbol{\Phi}(t) \tag{1.19}$$

For a small time interval t_0 to t we can assume $\frac{\partial \boldsymbol{f}(\boldsymbol{X},t)}{\partial \boldsymbol{X}(t)}$ to be constant and we can write

$$\boldsymbol{\Phi}(t) = \boldsymbol{\Phi}(t_0)e^{\frac{\partial \boldsymbol{f}(\boldsymbol{X},t)}{\partial \boldsymbol{X}(t)}(t-t_0)} \tag{1.20}$$

Note that by the definition of the state transition matrix, $\boldsymbol{\Phi}(t_0)$ is an identity matrix. Hence for a small time interval δt the state transition matrix at time t is

$$\boldsymbol{\Phi}(t) = e^{\frac{\partial \boldsymbol{f}(\boldsymbol{X},t)}{\partial \boldsymbol{X}(t)}\delta t} \tag{1.21}$$

1.3.2 Three-Dimensional Rotation Matrices

Rotation of vectors in three-dimensional space is essentially a linear transformation. this rotation can be performed using rotation matrices, which are also known as special orthogonal group SO(3). These rotation matrices are frequently used in navigation and tracking to change reference frames described in Chapter 2. The rotation matrices for rotation about the x, y, and z axes are listed below:

- The rotation matrix for counterclockwise rotation of a vector by an angle θ about the x-axis is given by:

$$\boldsymbol{R}_x(\theta) = \begin{bmatrix} 1 & 0 & 0 \\ 0 & \cos\theta & -\sin\theta \\ 0 & \sin\theta & \cos\theta \end{bmatrix} \tag{1.22}$$

- The rotation matrix for counterclockwise rotation of a vector by an angle θ about the y-axis is given by:

$$\boldsymbol{R}_y(\theta) = \begin{bmatrix} \cos\theta & 0 & \sin\theta \\ 0 & 1 & 0 \\ -\sin\theta & 0 & \cos\theta \end{bmatrix} \tag{1.23}$$

- The rotation matrix for counterclockwise rotation of a vector by an angle θ about the z-axis is given by:

$$\boldsymbol{R}_z(\theta) = \begin{bmatrix} \cos\theta & -\sin\theta & 0 \\ \sin\theta & \cos\theta & 0 \\ 0 & 0 & 1 \end{bmatrix} \tag{1.24}$$

For a rotation matrix $\boldsymbol{R}(\theta)$ the following properties hold:

1. $\boldsymbol{R}(\theta)^T = \boldsymbol{R}(\theta)^{-1}$
2. $\det(\boldsymbol{R}(\theta)) = 1$
3. $\boldsymbol{R}(-\theta) = \boldsymbol{R}(\theta)^{-1} = \boldsymbol{R}(\theta)^T$

1.3.3 Probability

As mentioned earlier, we pose the navigation and tracking problems as state estimation problems under uncertainty in the dynamics of space vehicles and uncertainty in the measurements. A solid understanding of probability and statistics is crucial for developing robust state estimation algorithms. In this section, we will cover the fundamental concepts of probability and statistics that are pertinent to state estimation in navigation and tracking in space.

1.3.3.1 Probability Space

A probability space is a mathematical construct that consists of three elements: a sample space, an event space, and a probability measure. It provides a formal framework for studying and analyzing random experiments and events. The three elements of a probability space are defined as follows:

- Sample Space (Ω): The sample space is the set of all possible outcomes of a random experiment. It is denoted by Ω and represents the entire set of possible outcomes or states that the experiment can result in. For example, if we toss a coin, the sample space is $\Omega = \{H, T\}$, where H represents heads and T represents tails. If we roll a die, the sample space is $\Omega = \{1, 2, 3, 4, 5, 6\}$, where each number represents the outcome of the die roll.

- Event Space ($\mathcal{F}$): The event space, also referred to as the σ-algebra or the set of events, is a collection of subsets of the sample space Ω. It contains the events or subsets that we are interested in studying or assigning probabilities to. The event space F satisfies certain properties, such as being closed under complementation, countable union, and countable intersection.
- Probability Measure (P): The probability measure is a function that assigns probabilities to events in the event space $\mathcal{F}$. It maps events to real numbers between 0 and 1, representing the likelihood or chance of their occurrence.

A probability space is often denoted as $(\Omega, \mathcal{F}, P)$.

1.3.3.2 Axioms of Probability

The probability measure mentioned above satisfies certain properties, known as the axioms of probability. The axioms of probability is also known as Kolmogorov's axioms. These axioms provide a mathematical framework for defining and analyzing probability. There are three axioms:

1. Nonnegativity: The probability of an event is a nonnegative real number. For any event A, the probability $P(A)$ is greater than or equal to zero: $P(A) \geq 0$.
2. Normalization: The total probability of the sample space Ω is equal to one; that is, $P(\Omega) = 1$.
3. Additivity: For any collection of mutually exclusive events (events that cannot occur simultaneously), the probability of the union of these events is equal to the sum of their individual probabilities. If $A_1, A_2, A_3, ..., A_n$ are mutually exclusive events, then the probability of their union is given by: $P(A_1 \cup A_2 \cup A_3 \cup ...) = P(A_1) + P(A_2) + P(A_3) + ... + P(A_n)$.

1.3.3.3 Conditional Probability

The conditional probability of an event A given another event B is defined as

$$P(A|B) = \frac{P(A \cap B)}{P(B)} \tag{1.25}$$

1.3.3.4 Independent Events

Two events A and B are said to be independent if

$$P(A \cap B) = P(A)P(B) \tag{1.26}$$

1.3.3.5 Random Variables

Consider a probability space $(\Omega, \mathcal{F}, P)$. A random variable X in this probability space is a function that maps each element in the sample space Ω to a real number:

$$X : \Omega \to \mathbb{R} \tag{1.27}$$

For each outcome $\omega \in \Omega$, $X(\omega)$ represents the numerical value associated with that outcome. The set of all possible numerical values that X can take is called the range or the codomain of the random variable. There are two main types of random variables:

- Discrete Random Variable: If the range of a random variable consists of a countable set of distinct values, it is called a discrete random variable. The probabilities associated with each value of the random variable can be represented using a probability mass function.
- Continuous Random Variable: If the range of a random variable is an interval or a continuous set of values, it is called a continuous random variable. The probabilities associated with intervals of values of the random variable are represented using a probability density function.

A multivariate random variable refers to a collection of two or more random variables that are defined on the same sample space. It represents a set of random quantities that are observed or measured simultaneously and may be dependent on each other. Essentially, a multivariate random variable is a vector of random variables.

1.3.3.6 Probability Mass Function

The probability mass function of a discrete random variable x that takes values from a discrete set $\zeta_1, \zeta_2, ..., \zeta_n$, is defined as

$$f_x(\zeta_i) = P(x = \zeta_i) \text{ for } i = 1, 2, ..., n \tag{1.28}$$

1.3.3.7 Probability Density Function

The probability density function of a continuous random variable $x \in \mathbb{R}$ at $x = \zeta$ is defined as

$$p(x = \zeta) = \lim_{d\zeta \to 0} \frac{P(\zeta - d\zeta < x < \zeta)}{d\zeta} > 0 \tag{1.29}$$

where

$$P(\alpha < x < \beta) = \int_{\alpha}^{\beta} p(x)dx \tag{1.30}$$

1.3.3.8 Expectation and Covariance of Random Variables

The expectation of a multivariate random variable $\boldsymbol{x} \in \mathbb{R}^n$ is defined as

$$E[\boldsymbol{x}] = \int_{-\infty}^{\infty} \boldsymbol{x}p(\boldsymbol{x})\boldsymbol{dx} = \bar{\boldsymbol{x}} \tag{1.31}$$

The covariance of a vector random variable $\boldsymbol{x}$ is defined as

$$\boldsymbol{\Sigma} = E[[\boldsymbol{x} - E[\boldsymbol{x}]][\boldsymbol{x} - E[\boldsymbol{x}]]^T] \tag{1.32}$$

1.3.3.9 Gaussian Distribution

A multivariate random variable $\boldsymbol{x}$ has a Gaussian distribution with mean $\boldsymbol{\mu}$ and covariance $\boldsymbol{\Sigma}$ if the probability distribution function is given by

$$p(\boldsymbol{x}; \boldsymbol{\mu}, \boldsymbol{\Sigma}) = \frac{1}{\sqrt{(2\pi)^n |\boldsymbol{\Sigma}|}} e^{-\frac{1}{2}(\boldsymbol{x}-\boldsymbol{\mu})^T \boldsymbol{\Sigma}^{-1}(\boldsymbol{x}-\boldsymbol{\mu})} \tag{1.33}$$

1.3.3.10 Stochastic Process

A stochastic process (also referred to as a random process) describes the evolution of a system over time in a probabilistic manner. It is a collection of random variables indexed by time, where each random variable represents the state of the system at a specific point in time.

Formally, a stochastic process can be defined as a family of random variables $X(t)$, $t \in T$, where T represents the index set that can be discrete or continuous, and $X(t)$ is the random variable representing the state of the process at time t. The index set T can be thought of as a time parameter, but it can also represent other quantities such as spatial locations or any other parameter of interest. The behavior of a stochastic process is characterized by its probability distribution or probability density function, which provides information about the likelihood of different values or trajectories of the process over time.

A stochastic process is called a discrete-time stochastic process if the time variable takes only positive integer values. A continuous-time stochastic process has a time variable that takes real values. For a system to be stochastic, one or more parts of the system have randomness associated with it. Unlike a deterministic system, for example, a stochastic system does not always produce the same output for a given input.

1.3.3.11 Markov Process

A stochastic process $\boldsymbol{X}_1, \boldsymbol{X}_2, \boldsymbol{X}_3, ...$ is called a Markov process if it has the following property:

$$P(\boldsymbol{X}_{n+1}|\boldsymbol{X}_n, \boldsymbol{X}_{n-1}, ..., \boldsymbol{X}_0) = P(\boldsymbol{X}_{n+1}|\boldsymbol{X}_n) \tag{1.34}$$

This property is called the Markov Property which essentially implies that the future state of the process only depends on the present state.

1.3.3.12 Process and Measurement Noise

The uncertainty in the system and the measurement is mathematically represented by process and measurement noise. In the space vehicle navigation context, system uncertainty arises due to various unmodeled forces that affect the position and velocity of the space vehicle. This uncertainty is modeled by incorporating a random variable term in the deterministic equation of motion and termed the process noise. Similarly, any physical measurement is uncertain; it contains a deterministic part and a random part. The randomness arises due to mechanical, thermal, and other external fluctuations. This random component is mathematically modeled by incorporating the measurement noise term in the measurement model equation. Throughout the book, we will consider the additive Gaussian process and measurement noise.

1.3.4 Motion of an Object in Space

Although we will describe the force model and the equations of motion in detail in the next chapter, we will briefly discuss the motion of an object in space in this section as preliminary. We will first describe Kepler's laws of planetary motion and then discuss the parameters that are used to describe the orbit of an object in space.

1.3.4.1 Kepler's Laws of Planetary Motion

Johannes Kepler, a German mathematician and astronomer, formulated three fundamental laws that describe the motion of planets and other celestial bodies in our solar system. These laws, known as Kepler's laws of planetary motion, were derived based on careful observations made by Kepler and previous astronomical data collected by Tycho Brahe. They played a pivotal role in the development of modern astronomy and our understanding of planetary motion. The three laws are as follows:

1. All planets move in elliptical orbits, with the Sun at one focus.
2. The line joining the planet to the Sun sweeps out equal areas in equal times as the planet travels around the ellipse.
3. The square of the orbital period of a planet is proportional to the cube of the semi-major axis of its orbit.

Kepler's laws provided crucial empirical evidence that influenced the development of Isaac Newton's laws of motion and universal gravitation. Newton was able to mathematically explain the underlying principles governing planetary motion by combining Kepler's laws with his own laws of motion and the law of universal gravitation. This resulted in a comprehensive understanding of celestial mechanics and marked a major milestone in the history of science.

1.3.4.2 Classical Orbital Elements

One standard and widely used method of describing orbit in space is through a set of six parameters which is also known as classical orbital elements or Keplerian elements. These parameters provide a concise mathematical representation of the shape, size, and orientation of an orbit.

The six orbital elements are:

- a semi major axis
- i inclination
- Ω right ascension (RA) of the ascending node

- e eccentricity
- ω argument of perigee
- ν true anomaly

The orbital elements are illustrated in Figure 1.2. These orbital elements are defined as follows:

1. Semi major axis (a) is half of the length of the major axis of an elliptical orbit.
2. Inclination (i) refers to the angle of the satellite's orbit plane with respect to the equatorial plane of the Earth. It is also the angle between the $\boldsymbol{Z}_{ECI}$ and the angular momentum vector $\boldsymbol{h}$. Mathematically [24],

$$\cos i = \frac{\boldsymbol{Z}_{ECI} \cdot \boldsymbol{h}}{|\boldsymbol{Z}_{ECI}|\,|\boldsymbol{h}|} \tag{1.35}$$

   Orbits with $0° < i < 90°$ are called prograde orbits in which the satellites move in the direction of the Earth's rotation. Orbits with $i = 90°$ are referred to as *polar orbits* in which the satellite travels over the poles. For $90° < i < 180°$ the orbit is called a *retrograde orbit* in which the satellite moves in the opposite direction of the Earth's rotation. For $i = 0°$ and $i = 180°$ the orbits are called *equatorial orbits*.
3. The right ascension of the ascending node (RAAN) Ω is measured counterclockwise in the equatorial plane from the vernal equinox direction to the ascending node (which is the point through which the orbit of the satellite passes through the plane of reference; i.e., equatorial plane from south to north direction).
4. The eccentricity vector, $\boldsymbol{e}$, points from the focus on the periapsis direction defining the shape of the conic orbit.
5. The argument of perigee ω is the angle between the eccentricity vector $\boldsymbol{e}$ and the node line vector shown in Figure 1.2, measured in the plane of the orbit in the direction of the satellite's motion.
6. True anomaly ν at an epoch is the sixth orbital element and it defines the angular location of the satellite in its orbit at epoch t relative to the location of periapsis [24].

The true anomaly θ is frequently replaced by the mean anomaly M (see Appendix A, Section 1 for details) [25]. An algorithm for computing the position vectors $\mathbf{r}$

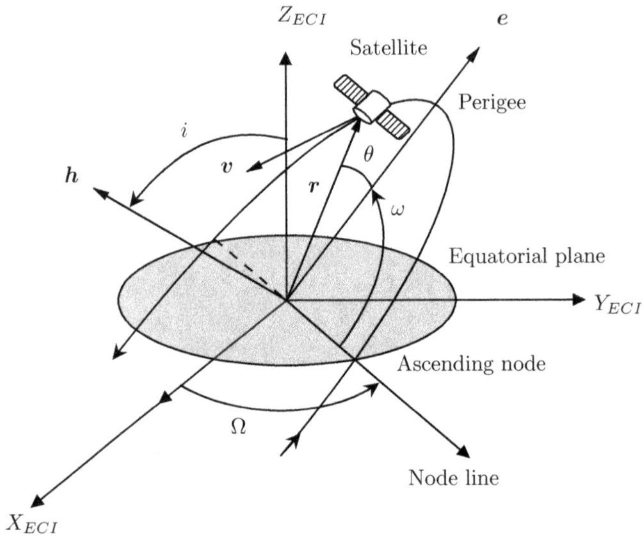

Figure 1.2: The orbital parameters in an Earth-centered inertial coordinate system.

and **v** from the orbital elements h, e, i, Ω, ω, and θ is given in Appendix A, Section 2. A reverse algorithm for computing the orbital elements from the position vectors is also given in Appendix A, Section 3. The detailed derivations of these algorithms can be found in [24].

References

[1] G. L. Smith, "Multivariable Linear Filter Theory Applied to Space Vehicle Guidance," *Journal of the Society for Industrial and Applied Mathematics Series a Control*, vol. 2, no. 1, pp. 19–32, Jan. 1964.

[2] H. Cox, "On the estimation of state variables and parameters for noisy dynamic systems," *IEEE Transactions on Automatic Control*, vol. 9, no. 1, pp. 5–12, Jan. 1964.

[3] M. Athans, R. Wishner, and A. Bertolini, "Suboptimal state estimation for continuous-time nonlinear systems from discrete noisy measurements," *IEEE Transactions on Automatic Control*, vol. 13, no. 5, pp. 504–514, Oct. 1968.

[4] B. Friedland and I. Bernstein, "Estimation of the state of a nonlinear process in the presence of nongaussian noise and disturbances," *Journal of the Franklin Institute*, vol. 281, no. 6, pp. 455–480, 1966.

[5] M. NøRgaard, N. K. Poulsen, and O. Ravn, "New developments in state estimation for nonlinear systems," *Automatica*, vol. 36, no. 11, pp. 1627–1638, 2000.

[6] S. Julier, "The spherical simplex unscented transformation," *Proceedings of the 2003 American Control Conference, 2003.*, vol. 3, Denver, CO, USA: IEEE, 2003, pp. 2430–2434.

[7] S. Julier and J. Uhlmann, "Unscented Filtering and Nonlinear Estimation," *Proceedings of the IEEE*, vol. 92, no. 3, pp. 401–422, Mar. 2004.

[8] N. Gordon, D. Salmond, and A. Smith, "Novel approach to nonlinear/non-Gaussian Bayesian state estimation," *IEE Proceedings F Radar and Signal Processing*, vol. 140, no. 2, p. 107, 1993.

[9] E. Jochim, E. Gill, O. Montenbruck, and M. Kirschner, "GPS based onboard and onground orbit operations for small satellites," *Acta Astronautica*, vol. 39, no. 9-12, pp. 917–922, Nov. 1996.

[10] J.-C. Yoon, B.-S. Lee, and K.-H. Choi, "Spacecraft orbit determination using GPS navigation solutions," *Aerospace Science and Technology*, vol. 4, no. 3, pp. 215–221, Apr. 2000.

[11] B.-S. Lee, J.-c. Yoon, Y. Hwang, and J. Kim, "Orbit determination system for the KOMPSAT-2 using GPS measurement data," *Acta Astronautica*, vol. 57, no. 9, pp. 747–753, Nov. 2005.

[12] O. Montenbruck, T. van Helleputte, R. Kroes, and E. Gill, "Reduced dynamic orbit determination using GPS code and carrier measurements," *Aerospace Science and Technology*, vol. 9, no. 3, pp. 261–271, Apr. 2005.

[13] O. Montenbruck, M. Garcia-Fernandez, A. Helm, and M. Markgraf, "Gps for microsatellites – status and perspectives," *In Proceedings of 6th IAA Symposium on Small Satellites for Earth Observation, April 23-26, Berlin*, 2007.

[14] S. Leung and O. Montenbruck, "Real-Time Navigation of Formation-Flying Spacecraft Using Global-Positioning-System Measurements," *Journal of Guidance, Control, and Dynamics*, vol. 28, no. 2, pp. 226–235, Mar. 2005.

[15] J. L. Crassidis and F. L. Markley, "Unscented Filtering for Spacecraft Attitude Estimation," *Journal of Guidance, Control, and Dynamics*, vol. 26, no. 4, pp. 536–542, Jul. 2003.

[16] J. Fisher and S. R. Vadali, "Gyroless Attitude Control of Multibody Satellites Using an Unscented Kalman Filter," *Journal of Guidance, Control, and Dynamics*, vol. 31, no. 1, pp. 245–251, Jan. 2008.

[17] J. Zhou, S. Knedlik, and O. Loffeld, "INS/GPS Tightly-Coupled Integration using Adaptive Unscented Particle Filter," *Journal of Navigation*, vol. 63, no. 03, pp. 491–511, Jul. 2010.

[18] W. Jiang, Y. Li, and C. Rizos, "Precise Indoor Positioning and Attitude Determination using Terrestrial Ranging Signals," *Journal of Navigation*, vol. 68, no. 02, pp. 274–290, Mar. 2015.

[19] W. F. Leven and A. D. Lanterman, "Unscented Kalman Filters for Multiple Target Tracking With Symmetric Measurement Equations," *IEEE Transactions on Automatic Control*, vol. 54, no. 2, pp. 370–375, Feb. 2009.

[20] S. Julier, J. Uhlmann, and H. Durrant-Whyte, "A new method for the nonlinear transformation of means and covariances in filters and estimators," *IEEE Transactions on Automatic Control*, vol. 45, no. 3, pp. 477–482, Mar. 2000.

[21] E.-J. Choi, J.-C. Yoon, B.-S. Lee, S.-Y. Park, and K.-H. Choi, "Onboard orbit determination using GPS observations based on the unscented Kalman filter," *Advances in Space Research*, vol. 46, no. 11, pp. 1440–1450, Dec. 2010.

[22] S. K. Biswas, L. Qiao, and A. G. Dempster, "A quantified approach of predicting suitability of using the Unscented Kalman Filter in a non-linear application," *Automatica*, vol. 122, p. 109 241, Dec. 2020.

[23] G. Strang, *Introduction to Linear Algebra*. SIAM, 2022.

[24] D. A. Vallado, *Fundamentals of Astrodynamics and Applications*. Springer Science & Business Media, 2001, vol. 12.

[25] H. D. Curtis, *Orbital Mechanics for Engineering Students* (Elsevier Aerospace Engineering Series), 1. ed., reprinted. Amsterdam: Elsevier/Butterworth-Heinemann, 2008.

Chapter 2

Space Vehicle Dynamics

2.1 OVERVIEW

From the navigation and tracking perspective, our interest is to estimate the position and velocity of a spacecraft as accurately as possible. This is a challenging task because the mathematical model of space vehicle dynamics is not exact, and the measurements are noisy. Here the differential equation that governs the time evolution of the position and velocity is referred to as the dynamics of the spacecraft. This is often referred to as the system model as well. This mathematical model of the spacecraft motion mismatches with the actual motion due to approximations and unmodeled dynamics. This holds true for any physical system. Control theorists have come up with an elegant way to account for these unmodeled dynamics and approximation in control and estimation problems by expressing the system dynamics using stochastic differential equations.

Following the conventional way of formulating the navigation and tracking problem, we will combine the position and velocity vectors of a spacecraft in a six-dimensional vector and refer to the state vector. The state vector is propagated using the spacecraft dynamics to compute the a priori estimate of the position and velocity of the spacecraft, which is required by a sequential estimator to estimate the a posteriori position and velocity from the observations or measurements. Essentially, a more accurate system model results in more accurate estimates of the position and velocity of the spacecraft. The accuracy requirement determines the level of the detailed model required in an estimation algorithm. In this chapter, we will discuss the construction of the mathematical models of the launch vehicle, reentry vehicle, and satellite dynamics.

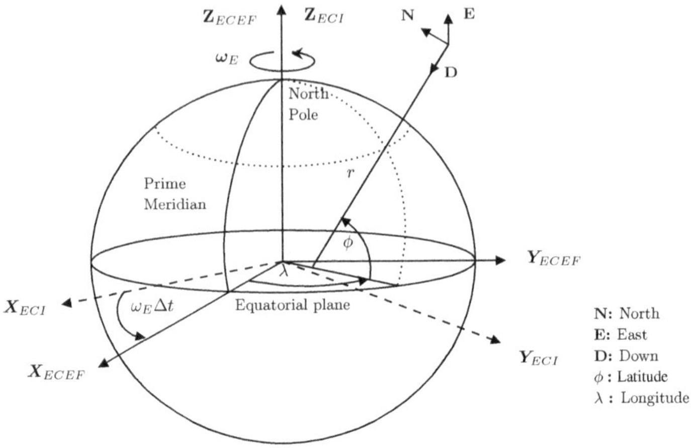

Figure 2.1: Reference frames.

2.2 REFERENCE FRAMES

Since the motion of a dynamic system is relative, a mathematical model of the motion must be expressed in a rigorously defined reference frame. A multitude of historical concepts in various reference frames is available. However, fundamentally there are two types of reference frames: inertial and noninertial. Defining an absolute inertial frame is a demanding task and not necessary most of the time. In practice, the conventional inertial frames are not inertial in a strict sense, because they are affected by the motion of the central body under consideration. Often these frames are referred to as quasi-inertial frames. Furthermore, noninertial frames are crucial because of the necessity of the knowledge of the position and velocity of a space vehicle relative to a particular position on the Earth, which is moving in three-dimensional space as the Earth rotates and follows a heliocentric orbit. Various reference frames are shown in Figure 2.1 and are described in the following sections.

2.2.1 Earth-Centered Inertial Frame

The origin of the earth-centered inertial (ECI) frame is the center of the Earth. In a strict sense, due to the motion of the Earth in space the ECI frame is a quasi-inertial frame with fixed directions of the *X*, *Y*, and *Z* axes. Various realizations of

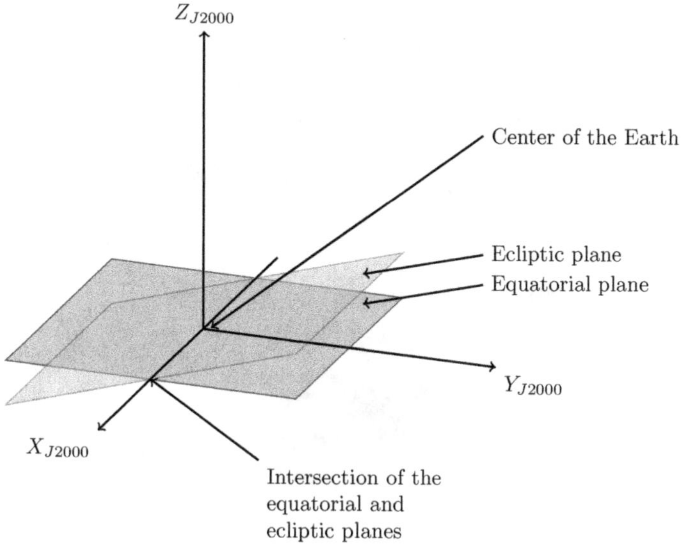

Figure 2.2: J2000 reference frame.

the ECI frame are available, however, the J2000 frame is the most commonly used ECI reference frame. In J2000 frame, as shown in Figure 2.2, the X-axis is defined in the direction of the intersection of the Earth's equatorial plane and the ecliptic plane at 12:00:00 on January 1, 2000 (J2000 epoch), and the Z-axis is normal to the mean equator at J2000 epoch. The Y-axis is defined by the cross-product of the X-axis and Z-axis directions. Another realization of the ECI frame is the geocentric celestial reference frame (GCRF) [1] and the axes of GCRF are closely aligned to the J2000 frame. The rotation between these two realizations is less than 0.1 arc second.

2.2.2 Earth-Centered Earth-Fixed Frame

The origin of the earth-centered earth-fixed (ECEF) frame is also the center of the Earth. The X-axis is aligned with the intersection of the equatorial plane and the prime meridian, the Z-axis is to the north pole (i.e., aligned with the axis of rotation of the Earth), and the Y-axis completes the right-hand rule [2]. The ECEF frame is a noninertial frame and continuously changes its orientation with respect to the ECI frame due to the Earth's rotation, precession, nutation, and polar motion. Earth's

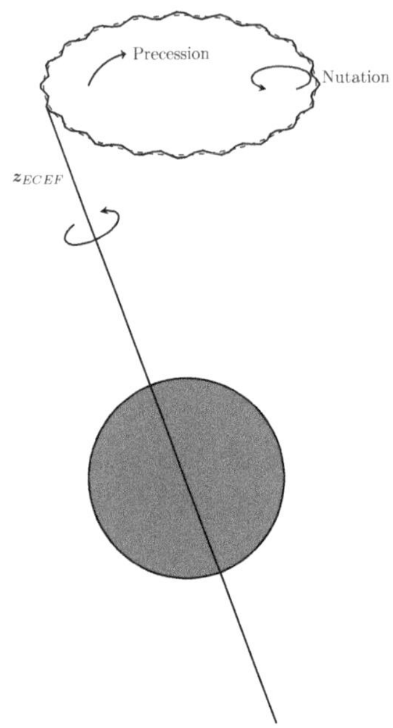

Figure 2.3: Precession and nutation of the Earth's axis of rotation.

precession is a slow, gradual periodic motion of the planet's axis of rotation. This motion is caused by the gravitational torque experienced by the nonspherical Earth due to the Sun and the Moon. The precession completes one full cycle every 26,000 years. Nutation is a small oscillation superimposed upon the precession motion caused by the monthly and annual variation of the lunar and solar gravitational torque. Consider $\boldsymbol{Z}_{ECEF}$ in Figure 2.3 as the axis of rotation of the Earth. The dashed circular path of the axis is due to the precession and the small oscillation on the circular path depicted by the solid curve is due to the nutation.

In addition to precession and nutation the Earth's axis changes its position with respect to the Earth's crust, which is referred to as polar motion. As a result, the conversion of a position vector $\boldsymbol{r}_{ECEF}$ from the ECEF frame to the ECI frame requires four rotational transformation matrices as shown in (2.1).

$$\boldsymbol{r}_{ECI} = \boldsymbol{T_P T_N T_\Omega T_{pm} r}_{ECEF} \tag{2.1}$$

Here, the transformation matrices $\boldsymbol{T}_P$, $\boldsymbol{T}_N$, $\boldsymbol{T}_\Omega$, and $\boldsymbol{T}_{pm}$ corresponds to precession, nutation, the Earth's rotation, and the polar motion, respectively. We will show a detailed transformation for a particular realization of the ECI and ECEF frames.

In the context of the ECEF reference frame, it is important to mention the World Geodetic System 84 (WGS84) reference system, which is a global standard for navigation and geospatial application. The origin of this reference system is the Earth's center of mass. The Z-axis of the WGS84 is the direction of the International Earth Rotation Service (IERS) reference pole [3], and the X-axis is defined by the intersection of the IERS reference meridian and the plane normal to the Z-axis and passing through the origin. The Y-axis is defined by the cross product of the X and Z-axes directions. WGS84 also defines the parameters for reference ellipsoid for the Earth, the nominal angular velocity and the geocentric gravitational constant for the Earth. These parameter values are provided in Table 2.1. Another commonly used realization of the ECEF frame is the International Terrestrial Reference Frame (ITRF). However, the difference between the WGS84 and the ITRF is less than 10 mm.

Table 2.1

WGS84 Parameters

Semi-Major Axis	6378137.0 m
Flattening Factor of the Earth	298.257223563
Nominal Mean Angular Velocity of the Earth	7292115×10^{-11} radian/s
Geocentric Gravitational Constant	$3.986004418 \times 10^{14}\ m^3/s^2$

Expressing the location on the Earth using polar coordinates (i.e., latitude, longitude, and altitude) is a common practice. Geodetic and geocentric coordinate systems can be used to define latitude, longitude, and altitude. In the geodetic coordinate system, the geodetic latitude is defined by the angle between the Earth's equatorial plane and the line normal to the reference ellipsoid at the location under consideration. The geocentric latitude is the angle between the equatorial plane and the line from the center of the Earth passing through the location. The geodetic and geocentric latitudes are shown in Figure 2.4. The ellipsoid in the figure represents the reference ellipsoid of the Earth and the horizontal line passing through the center of the ellipsoid (i.e., the center of mass of the Earth denotes the equatorial plane).

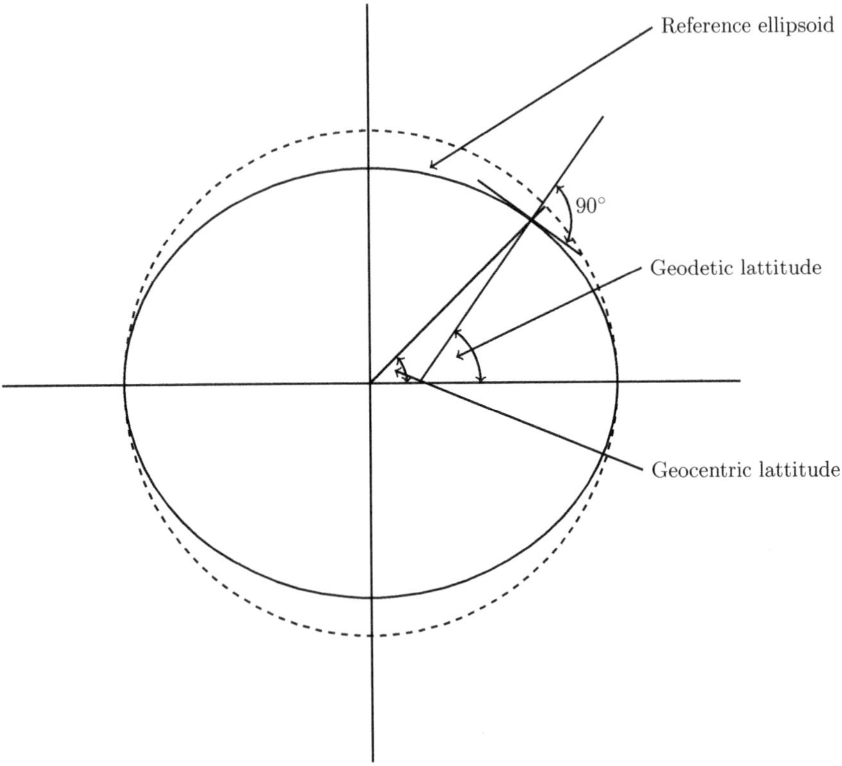

Figure 2.4: Geodetic and geocentric latitudes.

2.2.3 Topocentric Frame

The topocentric frame is another noninertial reference frame and is useful for ground-based measurements. For a given point on the Earth, the topocentric frame is aligned with the local horizontal plane. Generally, three orthogonal unit vectors point in the east, north, and zenith direction, respectively, to define the reference axes [4]. There can be various realizations of the topocentric frame based on which direction is considered as the x-axis. In Figure 2.1 the North-East-Down (NED) topocentric frame is shown where North is the x-axis, East is the y-axis, and Down is the z axis. The North-East-Up (NEU) topocentric frame is also commonly used where the z-axis is pointing upwards. The transformation between the ECEF and the topocentric frame can be easily performed using the latitude and longitude of the point of the Earth in the rotation matrices described in Chapter 1.

2.2.4 Solar Ecliptic Coordinate System

The solar ecliptic coordinate system (SE) is used in interplanetary trajectories. The *Z*-axis of the SE is normal to the ecliptic plane, the *X*-axis is towards the vernal equinox (conventionally at J2000 epoch), and the *Y*-axis can be defined by the cross product of the *X* and *Z* directions.

2.2.5 Reference Frame Conversion

Let us now consider particular realizations of the ECI and ECEF frames — GCRF and ITRF. The position $\boldsymbol{r}$, velocity $\boldsymbol{v}$, and acceleration $\boldsymbol{a}$ in GCRF are converted into ITRF using the following equations:

$$\boldsymbol{r}_{ITRF} = \boldsymbol{T}_{pm}^T \boldsymbol{T}_{\Omega}^T \boldsymbol{T}_N^T \boldsymbol{T}_P^T \boldsymbol{r}_{GCRF} \tag{2.2}$$

$$\boldsymbol{v}_{ITRF} = \boldsymbol{T}_{pm}^T \left[\boldsymbol{T}_{\Omega}^T \boldsymbol{T}_N^T \boldsymbol{T}_P^T \boldsymbol{v}_{GCRF} - \boldsymbol{\omega}_E \times \boldsymbol{r}_{PEF}\right] \tag{2.3}$$

$$\boldsymbol{a}_{ITRF} = \boldsymbol{T}_{pm}^T \left[\boldsymbol{T}_{\Omega}^T \boldsymbol{T}_N^T \boldsymbol{T}_P^T \boldsymbol{a}_{GCRF} - \boldsymbol{\omega}_E \times \boldsymbol{\omega}_E \times \boldsymbol{r}_{PEF} - 2\boldsymbol{\omega}_E \times \boldsymbol{v}_{PEF}\right] \tag{2.4}$$

Conversely, the position, velocity, and acceleration in ITRF can be converted using the equations provided below:

$$\boldsymbol{r}_{GCRF} = \boldsymbol{T}_{pm} \boldsymbol{T}_{\Omega} \boldsymbol{T}_N \boldsymbol{T}_P \boldsymbol{r}_{ITRF} \tag{2.5}$$

$$\boldsymbol{v}_{GCRF} = \boldsymbol{T}_{pm} \left[\boldsymbol{T}_{\Omega} \boldsymbol{T}_N \boldsymbol{T}_P \boldsymbol{v}_{ITRF} + \boldsymbol{\omega}_E \times \boldsymbol{r}_{PEF}\right] \tag{2.6}$$

$$\boldsymbol{a}_{GCRF} = \boldsymbol{T}_{pm}\left[\boldsymbol{T}_{\Omega}\boldsymbol{T}_{N}\boldsymbol{T}_{P}\boldsymbol{a}_{ITRF} + \boldsymbol{\omega}_{E}\times\boldsymbol{\omega}_{E}\times\boldsymbol{r}_{PEF} + 2\boldsymbol{\omega}_{E}\times\boldsymbol{v}_{PEF}\right] \quad (2.7)$$

In the above conversions, vectors with subscript PEF refer to the intermediate pseudo-Earth fixed (PEF) vectors. The polar motion matrix is defined as:

$$\begin{aligned}\boldsymbol{T}_{pm} &= \begin{bmatrix}1 & 0 & 0\\ 0 & \cos y_p & \sin y_p\\ 0 & -\sin y_p & \cos y_p\end{bmatrix}\begin{bmatrix}\cos x_p & 0 & -\sin x_p\\ 0 & 1 & 0\\ \sin x_p & 0 & \cos x_p\end{bmatrix}\\ &= \begin{bmatrix}\cos x_p & 0 & -\sin x_p\\ \sin x_p \sin y_p & \cos y_p & \cos x_p \sin y_p\\ \sin x_p \cos y_p & -\sin y_p & \cos x_p \cos y_p\end{bmatrix}\end{aligned} \quad (2.8)$$

Using the trigonometric small-angle approximation gives:

$$\boldsymbol{T}_{pm} = \begin{bmatrix}1 & 0 & -x_p\\ 0 & 1 & y_p\\ x_p & -y_p & 1\end{bmatrix} \quad (2.9)$$

The time-dependent components of polar motion are represented as

$$x_p = a_x + b_x t + c_{ax}\cos\left(\frac{2\pi t}{P_{ax}} + \phi_{ax}\right) + c_{cx}\cos\left(\frac{2\pi t}{P_{cx}} + \phi_{cx}\right) \quad (2.10)$$

$$y_p = a_y + b_y t + c_{ay}\cos\left(\frac{2\pi t}{P_{ay}} + \phi_{ay}\right) + c_{cy}\cos\left(\frac{2\pi t}{P_{cy}} + \phi_{cy}\right) \quad (2.11)$$

where parameters $a_x, a_y, b_x, b_y, c_{ax}, c_{ay}, c_{cx}, c_{cy}$ are obtained from a least-squares fit to six years of past polar motion data.

2.3 FORCE MODEL

The motion of a spacecraft can be described using Newton's laws of motion and the law of gravitation, which have been fundamental in understanding the behavior of moving objects for over 200 years. However, Albert Einstein expanded upon these classical theories with his theories of special and general relativity, which provided a more comprehensive framework for incorporating gravity into the realm of relativistic physics. The theory of general relativity accounts for several effects that classical gravitation cannot explain.

While the general relativistic equations are superior in describing the dynamics of a spacecraft, it is often computationally more efficient to utilize classical mechanics as a starting point for developing the spacecraft's motion dynamics. In this approach, the relativistic effects can be adjusted for in the measurement model, if necessary, making it more manageable from a state estimation perspective.

Let's begin by examining the gravitational force acting on a spacecraft. According to Newton's law of gravitation, when considering an artificial spacecraft of mass m positioned at $\boldsymbol{r}$ relative to the center of the Earth in an ECI frame, the force experienced by the satellite can be expressed as:

$$\boldsymbol{F} = -\frac{GMm}{r^3}\boldsymbol{r} \tag{2.12}$$

Here, G represents the universal gravitation constant and M denotes the mass of the Earth. The value of G is approximately $6.6743 \times 10^{-11} m^3 kg^{-1} s^{-2}$. It is important to note that this equation assumes the Earth is a perfect sphere with a uniform density. While this assumption is idealistic, we will later explore methods to account for deviations from this assumption. We will use the two-body problem to describe the motion of the spacecraft.

The two-body problem involves the interaction of two point masses in an isolated system, which interact through a central potential. In the context of orbital mechanics, this problem refers to the dynamics of a spacecraft under the influence of the gravitational force from a celestial body, typically the Earth. It is assumed that both bodies are spherically symmetric, and no external or internal forces act on the system except for the gravitational forces, which act along the line joining the centers of the two bodies [2].

The magnitude of the gravitational force F between two masses m_1 and m_2, with a distance r between their centers, can be given as:

$$F = G\frac{m_1 m_2}{r^2} \tag{2.13}$$

Let $\boldsymbol{r}$ be the position vector of m_2 relative to m_1. We can define $\boldsymbol{r}$ as $\boldsymbol{r}_2 - \boldsymbol{r}_1$ as shown in Figure 2.5. The gravitational force exerted on m_2 by m_1 can be written as:

$$\boldsymbol{F}_{21} = -\frac{Gm_1 m_2}{r^3}\boldsymbol{r} \tag{2.14}$$

The negative sign indicates that the force vector $\boldsymbol{F}_{21}$ is directed from m_2 towards m_1. Similarly, the force exerted on m_1 by m_2 can be given as:

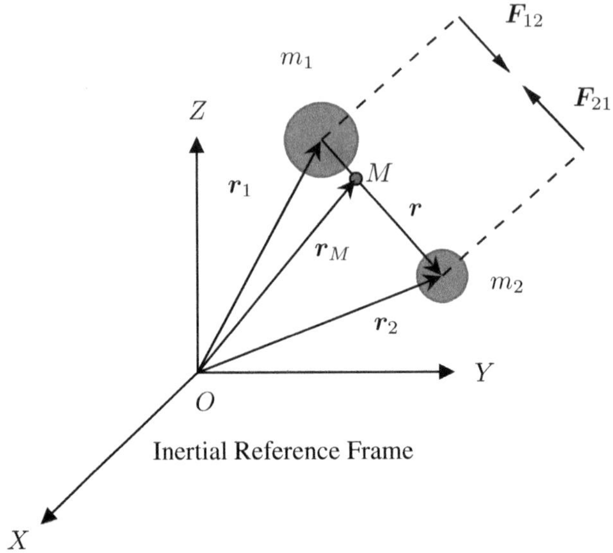

Figure 2.5: Two masses located in an inertial frame.

$$\boldsymbol{F}_{12} = \frac{Gm_1m_2}{r^3}\boldsymbol{r} \tag{2.15}$$

Now let's consider the equations of motion in an inertial frame for the two-body problem. The position vector of the center of mass M of the system is denoted as $\boldsymbol{R}_M$. It can be defined as:

$$\boldsymbol{R}_M = \frac{m_1\boldsymbol{R}_1 + m_2\boldsymbol{R}_2}{m_1 + m_2} \tag{2.16}$$

The absolute velocity $\boldsymbol{v}_M$ and absolute acceleration $\boldsymbol{a}_M$ of M are given by:

$$\boldsymbol{v}_M = \frac{m_1\dot{\boldsymbol{r}}_1 + m_2\dot{\boldsymbol{r}}_2}{m_1 + m_2} \tag{2.17}$$

$$\boldsymbol{a}_M = \frac{m_1\ddot{\boldsymbol{r}}_1 + m_2\ddot{\boldsymbol{r}}_2}{m_1 + m_2} \tag{2.18}$$

Applying Newton's second law of motion to body m_2, we have $\boldsymbol{F}_{21} = m_2\ddot{\boldsymbol{r}}_2$, where $\ddot{\boldsymbol{r}}_2$ represents the acceleration of m_2 measured relative to the ECI frame. Therefore, we can write:

$$-\frac{Gm_1m_2}{r^3}\boldsymbol{r} = m_2\ddot{\boldsymbol{r}}_2 \tag{2.19}$$

Now for m_1, we have:

$$\frac{Gm_1m_2}{r^3}\boldsymbol{r} = m_1\ddot{\boldsymbol{r}}_1 \tag{2.20}$$

By adding (2.19) and (2.20), we obtain:

$$m_1\ddot{\boldsymbol{r}}_1 + m_2\ddot{\boldsymbol{r}}_2 = 0 \tag{2.21}$$

This equation indicates that the acceleration of the center of mass M of the system is zero. Therefore, M moves with a constant velocity $\boldsymbol{v}_M$.

Now let's consider the equation of relative motion. By multiplying (2.19) by m_1 and (2.20) by m_2, we can obtain:

$$-\frac{Gm_1^2m_2}{r^3}\boldsymbol{r} = m_1m_2\ddot{\boldsymbol{r}}_2 \tag{2.22}$$

$$\frac{Gm_1m_2^2}{r^3}\boldsymbol{r} = m_1m_2\ddot{\boldsymbol{r}}_1 \tag{2.23}$$

By subtracting (2.23) from (2.22), we can derive the second-order differential equation of the two-body problem:

$$\ddot{\boldsymbol{r}} = -\frac{G(m_1 + m_2)}{r^3}\boldsymbol{r} \tag{2.24}$$

For the system consisting of the Earth and a spacecraft with masses M and m respectively, $M \gg m$. Then in the above equation, $G(M + m) \approx GM$. Let us define $\mu = GM$, which is referred as the gravitational parameter of the Earth. The value of μ is 3.986004418 $\times$ $10^{-14}m^3s^{-2}$. Note that the center of the mass of this system is also approximately the center of mass of the Earth because $m_1 \gg m_2$.

By using the gravitational parameter μ in (2.24), we can express the second-order differential equation for the two-body problem as:

$$\ddot{\boldsymbol{r}} + \frac{\mu}{r^3}\boldsymbol{r} = \boldsymbol{0} \tag{2.25}$$

Equation (2.25) is a fundamental equation used to model the motion of satellites or spacecraft in the vicinity of a celestial body such as the Earth.

2.3.1 Perturbation Forces

As discussed in the previous section, (2.25) describes the motion of a spacecraft orbiting around the Earth in an ideal scenario. Perturbation forces represent the departure from this ideal model and account for the effect of the nonuniform shape and density of the Earth, atmospheric drag, solar radiation pressure, and the gravitational force due to other celestial bodies on the spacecraft. Fortunately, our physical realization of force is additive in nature, and we can address these perturbation forces by adding an acceleration term $\boldsymbol{a}$ to our existing equation (2.25):

$$\ddot{\boldsymbol{r}} = -\frac{GM}{r^3}\boldsymbol{r} + \boldsymbol{a} \tag{2.26}$$

where

$$\boldsymbol{a} = \boldsymbol{a}_g + \boldsymbol{a}_d + \boldsymbol{a}_s + \boldsymbol{a}_c \tag{2.27}$$

and $\boldsymbol{a}_g, \boldsymbol{a}_d, \boldsymbol{a}_s, \boldsymbol{a}_c$ are perturbations due to the nonuniform shape and density of the Earth, the atmospheric drag, solar radiation pressure, and the gravitational force due to other celestial bodies, respectively. Equation (2.26) is known as Cowell's formulation.

2.3.1.1 Nonuniform Gravitational Field of the Earth

Considering the nonuniform shape and density, the Earth's gravity potential in the ECEF frame can be defined as [1]

$$U = \frac{\mu}{r} + U' \tag{2.28}$$

where

$$U' = \frac{mu}{r}\sum_{l=1}^{\infty}\sum_{m=0}^{l}\left(\frac{R_e}{r}\right)^l P_{lm}(\sin\phi)[C_{lm}\cos m\lambda + S_{lm}\sin m\lambda] \tag{2.29}$$

and R_e is the Earth's mean equatorial radius, P_{lm} is the associated Legendre function of degree and order l and m, ϕ is the geocentric latitude, and λ is the longitude of the spacecraft. C_{lm} and S_{lm} are the mass coefficients of the Earth. Terms with $m = 0$ have no dependence on the longitude and corresponding coefficients C_{l0} are called zonal coefficients. Coefficients for $l \neq m$ and $l = m$ are called tesseral and sectoral coefficients, respectively. It should be noted that these coefficients cannot be calculated using mathematical formulations. The mass coefficients are determined indirectly by satellite tracking, surface gravimetry, and altimetry [5]. This mass coefficient information is referred to as the gravity model. A standard gravity model is the Earth Gravity Model 2008 (EGM2008) published by the National Geospatial-Intelligence Agency (NGA) of the United States. The acceleration due to the Earth's gravity potential U

$$\ddot{\boldsymbol{r}} = \boldsymbol{\nabla} U \tag{2.30}$$

The associated Legendre function:

$$P_{lm}(\sin\phi) = (\cos\phi)^m \frac{d^m}{d\,(\sin\phi)^m}\left[P_l(\sin\phi)\right] \tag{2.31}$$

with the Legendre polynomial:

$$P_l(\sin\phi) = \frac{1}{2^l l!}\frac{d^l}{d(\sin\phi)^l}(\sin^2\phi - 1)^l \tag{2.32}$$

The associated Legendre function and the mass coefficients are generally normalized for the ease of implementation, because, without normalization, the mass coefficients span 10 or more orders of magnitude [4]. The normalized associated Legendre function is

$$\overline{P}_{lm}(\sin\phi) = \sqrt{\frac{(l-m)!k(2l+1)!}{(l+m)!}}P_{lm}(\sin\phi) \tag{2.33}$$

and the normalized mass coefficients:

$$\begin{bmatrix}\overline{C}_{lm}\\ \overline{S}_{lm}\end{bmatrix} = \sqrt{\frac{(l+m)!}{(l-m)!k(2l+1)!}}\begin{bmatrix}C_{lm}\\ S_{lm}\end{bmatrix} \tag{2.34}$$

where

$$\begin{aligned} k &= 1; \mathit{if}\, m = 0 \\ k &= 2; \mathit{if}\, m \neq 0 \end{aligned} \tag{2.35}$$

Then the gravitational potential,

$$\begin{aligned} &U(r,\phi,\lambda) \\ &= \frac{\mu}{r} + \mu\sum_{l=1}^{\infty}\sum_{m=0}^{l}\left(\frac{R_E}{r}\right)^l \overline{P}_{lm}(\sin\phi)\left(\overline{C}_{lm}\cos(m\lambda) + \overline{S}_{lm}\sin(m\lambda)\right) \end{aligned} \tag{2.36}$$

Note that U', the perturbation term in the gravitational potential is a function of r, ϕ, and λ and hence defined in the ECEF frame. Then the acceleration due to this perturbation in the ECEF frame is

$$\ddot{\boldsymbol{r}}'_{ECEF} = \begin{bmatrix} \frac{\partial r}{\partial x'} & \frac{\partial \phi}{\partial x'} & \frac{\partial \lambda}{\partial x'} \\ \frac{\partial r}{\partial y'} & \frac{\partial \phi}{\partial y'} & \frac{\partial \lambda}{\partial y'} \\ \frac{\partial r}{\partial z'} & \frac{\partial \phi}{\partial z'} & \frac{\partial \lambda}{\partial z'} \end{bmatrix} \begin{bmatrix} \frac{\partial U'}{\partial r}, \\ \frac{\partial U'}{\partial \phi} \\ \frac{\partial U'}{\partial \lambda} \end{bmatrix} \tag{2.37}$$

where x', y', and z' are in the spacecraft coordinate in the ECEF frame and

$$\frac{\partial U'}{\partial r} = -\frac{\mu}{r^2}\left[\sum_{l=1}^{\infty}(l+1)\left(\frac{R_E}{r}\right)^l \sum_{m=0}^{l} \overline{P}_{lm}(\sin\phi)\left(\overline{C}_{lm}\cos(m\lambda) + \overline{S}_{lm}\sin(m\lambda)\right)\right] \tag{2.38}$$

$$\frac{\partial U'}{\partial \phi} = \frac{\mu}{r}\sum_{l=1}^{\infty}\left(\frac{R_E}{r}\right)^l \sum_{m=0}^{l} \cos\phi \frac{d\left(\overline{P}_{lm}(\sin\phi)\right)}{d(\sin\phi)}\left(\overline{C}_{lm}\cos(m\lambda) + \overline{S}_{lm}\sin(m\lambda)\right) \tag{2.39}$$

$$\frac{\partial U'}{\partial \lambda} = \frac{\mu}{r}\sum_{l=1}^{\infty}\left(\frac{R_E}{r}\right)^n \sum_{m=0}^{l} \overline{P}_{lm}(\sin\phi) m\left(-\overline{C}_{lm}\sin(m\lambda) + \overline{S}_{lm}\cos(m\lambda)\right) \tag{2.40}$$

The matrix in (2.37) is the transformation matrix from spherical to ECEF coordinate. It can be shown that [6]

$$\begin{bmatrix} \frac{\partial r}{\partial x'} & \frac{\partial \phi}{\partial x'} & \frac{\partial \lambda}{\partial x'} \\ \frac{\partial r}{\partial y'} & \frac{\partial \phi}{\partial y'} & \frac{\partial \lambda}{\partial y'} \\ \frac{\partial r}{\partial z'} & \frac{\partial \phi}{\partial z'} & \frac{\partial \lambda}{\partial z'} \end{bmatrix} = \begin{bmatrix} \cos\phi\cos\lambda & -\sin\phi\cos\lambda & -\sin\lambda \\ \cos\phi\sin\lambda & -\sin\phi\sin\lambda & \cos\lambda \\ \sin\phi & \cos\phi & 0 \end{bmatrix} \tag{2.41}$$

Let us denote this matrix as $\boldsymbol{T}_{r\phi\lambda}^{xyz}$. If the gradient of U is taken in a nonrotating inertial frame to obtain the acceleration of the spacecraft becomes [6]

$$\ddot{\boldsymbol{r}} = -\frac{GM}{r^3}\boldsymbol{r} + \boldsymbol{T}_{xyz}^{XYZ}\boldsymbol{T}_{r\phi\lambda}^{xyz}\nabla U' \tag{2.42}$$

where $\boldsymbol{T}_{r\phi\lambda}^{xyz}$ is the transformation matrix from spherical to ECEF coordinates, and $\boldsymbol{T}_{xyz}^{XYZ}$ is the transformation matrix from ECEF to ECI coordinates. Note that the gradient of U' in the spherical coordinate has singularities at poles. This can be avoided using Pine's formulation [7] where a Cartesian coordinate system is used without the requirement of changing the spherical harmonics coefficients. Lundberg and Schutz suggested recursive computation of derived Legendre functions to obtain a numerically stable and precise solution using Pine's formulation [8].

2.3.1.2 Atmospheric Drag

A widely accepted model of the acceleration due to atmospheric drag is [9]

$$\boldsymbol{a}_d = -\frac{1}{2}C_D\frac{A}{m}\rho v_r^2\hat{\boldsymbol{u}}_v \tag{2.43}$$

which considers spherical spacecraft. Here C_D is the drag coefficient of the spacecraft (range: $1.5 - 3.0$), A is the cross-section area of the spacecraft, m is the mass of a satellite, ρ is the atmospheric density at the satellite's altitude, and v_r is the relative velocity of the spacecraft with respect to the atmosphere and $\hat{\boldsymbol{u}}_v = \frac{\boldsymbol{v}_r}{v_r}$. $\boldsymbol{v}_r$ can be defined as

$$\boldsymbol{v}_r = \boldsymbol{v} - \boldsymbol{v}_{atm} \tag{2.44}$$

where $\boldsymbol{v}_{atm}$ is the velocity of the atmosphere that is unknown [9]. C_D depends on the shape of the space vehicle as well as altitude [10]. For a specific satellite, the two line element (TLE) contains a drag like parameter B^* which is defined as

$$B^* = \frac{1}{2}\frac{C_D A}{m}\rho_0 \tag{2.45}$$

where NOARD defines the atmospheric density at the perigee of the orbit: $\rho_0 = 2.461 \times 10^{-5} kg/m^3$. Hence one can compute C_D from the B^* coefficient. Generally, C_D is treated as a random variable and estimated through the navigation filter. It should be noted that the cross-sectional area of the satellite also varies due to the change of attitude as most of the satellites are not spherical. For nonspherical shapes, the attitude solution can be used to compute the effective area for the calculation of the atmospheric drag [9]. An extensive analysis of other different methods, for example, Monte Carlo-based and computational fluid dynamics (CFD)-based approaches for the atmospheric drag model, has been discussed in [11] and [9].

It should be noted that the atmospheric density depends on the altitude. There are various models available to compute the density of the atmosphere. The Jacchia-Roberts Atmosphere Model is one of the widely used models [4], which is a time-varying model. The Jacchia-Roberts model is known for its accuracy, but it can be computationally heavy due to its complexity. It takes into account various factors such as solar flux, geomagnetic activity, and time of year to provide accurate density estimates.

Alternatively, a simpler approach is to use an exponential model to determine the density of the upper atmosphere. This model assumes an exponential decrease in density with increasing altitude. While it may not capture the detailed variations of the atmosphere as accurately as the Jacchia-Roberts model, the exponential model offers a computationally efficient way to estimate the density at higher altitudes. It is often used for quick approximations or preliminary calculations where high accuracy is not critical. In this static model the atmospheric density ρ decays exponentially with increasing altitude h between 0—1000 km [1]:

$$\rho = \rho_0 e^{-\frac{h_e - h}{H}} \tag{2.46}$$

where, ρ_0 is the reference density, h is the reference altitude, h_e is the actual altitude above the reference ellipsoid, and H = the scale height.

2.3.1.3 Solar Radiation Pressure

A spacecraft experiences a force induced by the transfer of momentum through impact, reflection, or absorption of photons from the Sun. This is known as solar radiation pressure, and the acceleration due to the solar radiation pressure on a spacecraft is given by [4]

$$\boldsymbol{a}_s = -P_S C_R \frac{A}{m} \frac{\boldsymbol{r}_s}{r_s^3} \tag{2.47}$$

where P_S is the solar radiation pressure ($\approx 4.56 \times 10^{-6} N.m^{-2}$), and C_R is the solar radiation pressure coefficient of the satellite. $C_R = 1 + \varepsilon$, ε is the reflectivity of the selected satellite component, and $\boldsymbol{r}_s$ is the position vector of the satellite with respect to the Sun in the ECI frame. In this formulation, A denotes the surface area exposed to solar illumination. Consequently, A depends on the satellite's attitude with respect to the Sun. Similar to C_D, C_R is generally included in the state vector as a state variable and considered as a parameter to be estimated.

This model is a simplified approach to solar radiation pressure. A more accurate model considering the eclipse condition is given in [4] which is used for high-precision applications.

2.3.1.4 Gravitational Force Due to Other Celestial Objects

A spacecraft's motion is influenced by the gravitational force of multiple celestial objects other than the Earth. Essentially the equation of motion of spacecraft becomes a many-body problem, and it is well-known that there is no analytical solution to the many-body problem. However, one can find a numerical solution to the many-body problem using Cowell's formulation, as mentioned in Section 2.3.1. The acceleration due to the gravitational force of celestial objects other than the central body is given by

$$\boldsymbol{a}_c = \sum_{j=1}^{n_p} GM_j \left(\frac{\boldsymbol{r}_{js}}{r_{js}^3} - \frac{\boldsymbol{r}_j}{{r_j}^3} \right) \tag{2.48}$$

where $\boldsymbol{r}_{js}$ is the position vector of a spacecraft with respect to the j-th celestial object, and $\boldsymbol{r}_j$ is the position vector of the j-th celestial object with respect to the central body.

Let us consider the case of a satellite orbiting the Earth and consider the Sun and the Moon as the perturbing bodies. We will examine the perturbing acceleration due to the Sun and the Moon on the satellite. Let the position of the satellite, the Earth, the Sun, and the Moon be $\boldsymbol{r}_{sat}$, $\boldsymbol{r}_E$, $\boldsymbol{r}_S$, and $\boldsymbol{r}_M$, respectively, with respect to an inertial frame. The satellite position with respect to the Earth is $\boldsymbol{r}_{Esat} = \boldsymbol{r}_{sat} - \boldsymbol{r}_E$, the satellite position with respect to the Sun is $\boldsymbol{r}_{Ssat} = \boldsymbol{r}_{sat} - \boldsymbol{r}_S$, and the satellite position with respect to the Moon is $\boldsymbol{r}_{Msat} = \boldsymbol{r}_{sat} - \boldsymbol{r}_M$. Additionally the position the Sun with respect to the Earth is $\boldsymbol{r}_{ES} = \boldsymbol{r}_S - \boldsymbol{r}_E$ and the position of the Moon with respect to the Earth is $\boldsymbol{r}_{EM} = \boldsymbol{r}_M - \boldsymbol{r}_E$. The geometry is shown in Figure 2.6.

Applying Newton's second law of motion and the law of gravitation to the satellite and Earth, we obtain the sum of the forces acting on the Earth due to the combined effect of the satellite pulling on the Earth and the Sun pulling on the Earth is

$$M\ddot{\boldsymbol{r}}_E = \frac{GMm\boldsymbol{r}}{r^3} + \frac{GMM_s\boldsymbol{r}_{ES}}{r_{ES}^3} \tag{2.49}$$

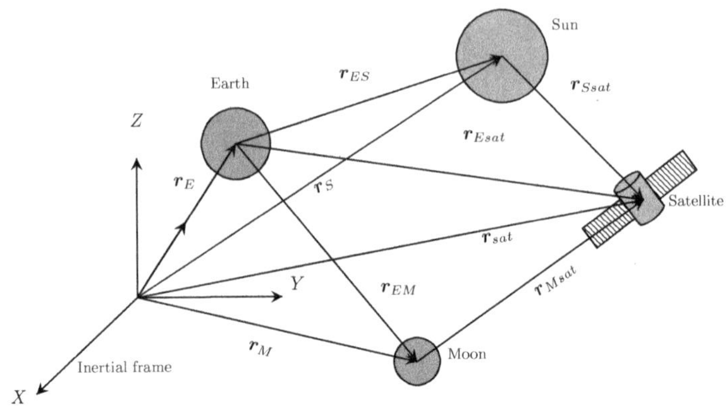

Figure 2.6: Satellite under the influence of the gravitation of the Earth, the Moon, and the Sun.

The sum of the forces acting on the satellite becomes:

$$m\ddot{\boldsymbol{r}}_{sat} = -\frac{GMm\boldsymbol{r}}{r^3} - \frac{GM_s m\boldsymbol{r}_{Ssat}}{r^3_{Ssat}} \tag{2.50}$$

The second time derivative of $\boldsymbol{r} = \boldsymbol{r}_{sat} - \boldsymbol{r}_E$ yields

$$\ddot{\boldsymbol{r}} = \ddot{\boldsymbol{r}}_{sat} - \ddot{\boldsymbol{r}}_E \tag{2.51}$$

By subtracting (2.50) and (2.49) and using (2.51) we get

$$\ddot{\boldsymbol{r}} = -\frac{GM\boldsymbol{r}}{r^3} - \frac{GM_s\boldsymbol{r}_{Ssat}}{r^3_{Ssat}} - \frac{Gm\boldsymbol{r}}{r^3} - \frac{GM_s\boldsymbol{r}_{ES}}{r^3_{ES}}$$

Grouping like terms and using $\ddot{\boldsymbol{r}}_{satS} = -\ddot{\boldsymbol{r}}_{Ssat}$, we obtain

$$\ddot{\boldsymbol{r}} = -\frac{(GM+m)\boldsymbol{r}}{r^3} + GM_s\left(\frac{\boldsymbol{r}_{satS}}{r^3_{satS}} - \frac{\boldsymbol{r}_{ES}}{r^3_{ES}}\right) \tag{2.52}$$

The first term is the acceleration on the satellite due to a point mass central body (i.e., the Earth). The second term is called the direct term depending on the distance to the third body from the satellite. The third term is called the indirect term, which depends on the position of the third body relative to the centered body (i.e., the

Earth). Similarly, considering the Sun and the Moon as the perturbing body, we obtain the expression of the Lunisolar gravitational acceleration in the ECI frame:

$$\ddot{\boldsymbol{r}} = -\mu\frac{\boldsymbol{r}}{r^3} + \mu_S\left(\frac{\boldsymbol{r}_{satS}}{r_{satS}^3} - \frac{\boldsymbol{r}_{ES}}{r_{ES}^3}\right) + \mu_M\left(\frac{\boldsymbol{r}_{satM}}{r_{satM}^3} - \frac{\boldsymbol{r}_{EM}}{r_{EM}^3}\right) \tag{2.53}$$

where μ_S and μ_M are the gravitational parameters of the Sun and the Moon respectively. The perturbation acceleration due to solar gravity ($\boldsymbol{a}_S$) and lunar gravity ($\boldsymbol{a}_M$) are

$$\boldsymbol{a}_s = \mu_S\left(\frac{\boldsymbol{r}_{satS}}{r_{satS}^3} - \frac{\boldsymbol{r}_{ES}}{r_{ES}^3}\right) \tag{2.54}$$

and

$$\boldsymbol{a}_M = \mu_M\left(\frac{\boldsymbol{r}_{satM}}{r_{satM}^3} - \frac{\boldsymbol{r}_{EM}}{r_{EM}^3}\right) \tag{2.55}$$

2.3.2 Thrust

In addition to the forces described in previous sections, launch vehicles as well as other spacecraft experience forces in terms of thrust. The sources of these thrusts are rocket motors. Rocket motors fixed with launch vehicles are the primary source of thrust that is required for the ascent of the launch vehicle and the containing payload spacecraft. Rocket motors that are fixed with spacecraft are often referred to as thrusters, and these are responsible for station keeping, attitude control, mid-course correction, and so forth. The thrust generated by these rocket motors or thrusters can be expressed as

$$T = \dot{m}_e v_e \tag{2.56}$$

where $\dot{m}_e$ is the exhaust mass flow rate and v_e is the effective exhaust velocity. Note that this is the magnitude of the thrust. The direction is aligned with the direction of the exhaust nozzle of a rocket motor.

2.4 STATE SPACE MODEL OF SPACE VEHICLES

It should be noted that the construction of mathematical models for various forces is not enough to solve navigation and tracking problems. We need to express the dynamics in the state space form. The state space model essentially breaks a n-th order differential equation to n first order differential equations and expresses them in a compact n-dimensional vector differential equation. As mentioned in previous sections, C_D and C_R are often estimated with the position and velocity of the space vehicle and hence the state space model needs to be constructed accordingly.

Essentially, the acceleration models for all space vehicles are similar and can be constructed using the force models discussed in Section 2.3. However, the state space models can be constructed differently for different types of space vehicles based on the choice of state variables. For example, it is possible to construct the state space model of a satellite considering Keplerian elements — orbital inclination, right ascension angle of the ascending node, eccentricity, semimajor axis, argument of perigee, and true anomaly as state variables instead of position and velocity vector. Similarly, for the launch and reentry vehicles one can use position and velocity vectors in a suitable Cartesian coordinate system to develop the state space models. In this book, we have selected position and velocity to represent the state of satellites and interplanetary spacecraft, whereas, for the launch and reentry vehicles, we have selected altitude, down-range, longitudinal speed, and some other variables in the state vector to construct simple models. One should choose the state variables considering the mission, application, and compatibility with legacy software. In the rest of the section, we will discuss the construction of state space models for a range of space vehicles.

2.4.1 Earth Orbiting Satellites

Let the position vector of a satellite be $\boldsymbol{r}$ and the velocity is $\boldsymbol{v}$. For simplicity, let us neglect all the perturbation forces and consider the Earth as a perfect sphere with uniform density. Under these assumptions, we can use (2.12) and write

$$\ddot{\boldsymbol{r}} = \dot{\boldsymbol{v}} = -\frac{\mu}{r^3}\boldsymbol{r} \tag{2.57}$$

Also, notice that

$$\dot{\boldsymbol{r}} = \boldsymbol{v} \tag{2.58}$$

Combining (2.58) and (2.57) one can write

$$\begin{bmatrix} \dot{\boldsymbol{r}} \\ \dot{\boldsymbol{v}} \end{bmatrix} = \begin{bmatrix} \boldsymbol{v} \\ -\frac{\mu}{r^3}\boldsymbol{r} \end{bmatrix} \tag{2.59}$$

One can always include the acceleration due to perturbation forces in (2.59).

2.4.2 Launch Vehicle

A launch vehicle trajectory, in reality, is three-dimensional. In addition, the attitude of the launch vehicle tends to change during the ascent phase. While tracking the

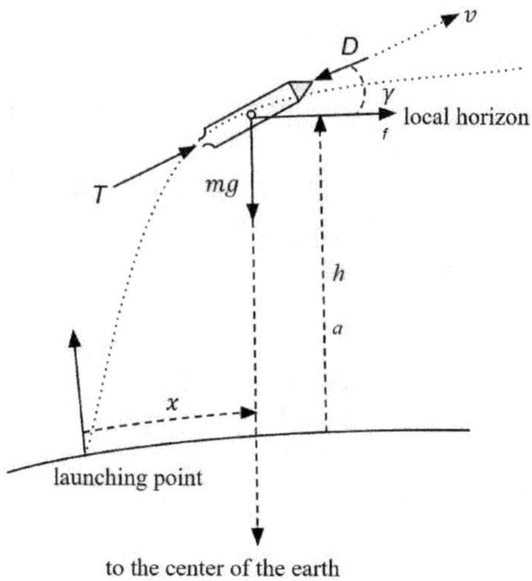

Figure 2.7: Launch vehicle trajectory.

launch vehicle, the attitude is also determined in practice. Attitude determination and navigation or tracking can be done using a single estimator; however, the dimensionality of the problem increases. Generally, these two problems are solved separately. The trajectory of a launch vehicle is shown in Figure 2.7. As shown in the figure, the thrust T acts in the direction of the longitudinal velocity and the atmospheric drag D acts in the opposite direction. The angle between the velocity and the local horizon is the flight path angle γ_f.

For simplicity of analysis, let's consider the trajectory of a launch vehicle two-dimensional. In that case, we are interested in estimating the down-range distance x, altitude h_a, and longitudinal speed v. However, these variables depend on other parameters: the flight path angle of the launch vehicle γ_f, aerodynamic coefficient C_D and mass m. Including all these variables and parameters in the state vector, the state space model of a launch vehicle can be written as:

$$\begin{bmatrix} \dot{x} \\ \dot{h}_a \\ \dot{v} \\ \dot{\gamma}_f \\ \dot{m} \\ \dot{C}_D \end{bmatrix} = \begin{bmatrix} \frac{R_E}{R_E+h_a} v \cos\gamma_f \\ v \sin\gamma_f \\ \frac{T}{m} - \frac{D}{m} - g\sin\gamma_f \\ -\frac{1}{v}\left(g - \frac{v^2}{R_E+h_a}\right)\cos\gamma_f \\ -\dot{m}_e \\ 0 \end{bmatrix} \tag{2.60}$$

where T is the engine thrust, D is the aerodynamic drag, g is the gravitational acceleration, R_E is the local radius of the Earth, and $\dot{m}_e$ is the exhaust mass flow rate. These parameters depend on the launch vehicle construction and the mission requirement. The drag force is modeled using an exponential atmospheric density model [12]. The drag force equation is [13]

$$D = \frac{1}{2} A C \rho_0 e^{-\frac{h_a}{H}} v^2 \tag{2.61}$$

where A is the frontal area of the launch vehicle, ρ_0 is the atmospheric density at the sea level, and $H = 7.1628$ km [12] is the scale height.

2.4.3 Reentry Vehicle

The reentry vehicle follows a curved path during atmospheric reentry. In detailed reentry dynamics, the downrange x, altitude h, velocity v, flightpath angle γ_f and the aerodynamic coefficient C_D must be considered as state variables. A planar reentry vehicle trajectory and the direction of various forces are provided in Figure 2.8. The reentry dynamics can be expressed as [14]:

$$\begin{bmatrix} \dot{x} \\ \dot{h}_a \\ \dot{v} \\ \dot{\gamma}_f \\ \dot{C}_D \end{bmatrix} = \begin{bmatrix} \frac{R_E}{R_E+h_a} v \cos\gamma_f \\ v \sin\gamma_f \\ -\frac{D}{m} - g\sin\gamma_f \\ -\frac{1}{v}\left(g - \frac{v^2}{R_E+h_a}\right)\cos\gamma_f \\ 0 \end{bmatrix} \tag{2.62}$$

where m is the mass of the reentry vehicle, R_E is the mean radius of the Earth, g is gravitational acceleration, D is the aerodynamic drag. Similar to the launch vehicle model, D is modeled using (2.61).

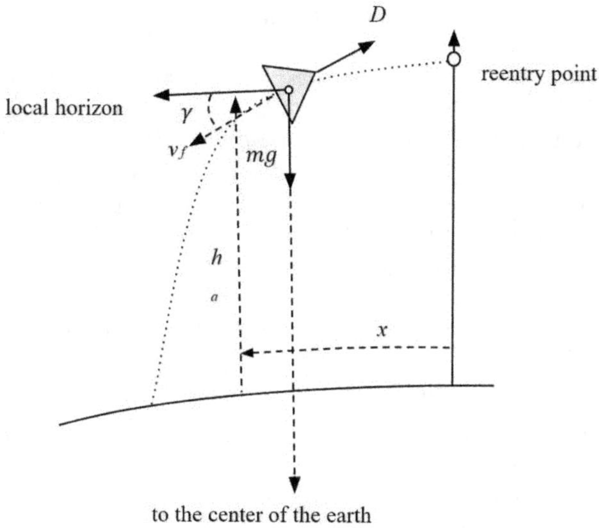

Figure 2.8: Reentry vehicle trajectory.

2.4.4 Interplanetary Spacecraft

Interplanetary spacecraft are sometimes parked in an earth orbit before entering the interplanetary trajectory. When in an earth orbit, the state space model described in Section 2.4.1 can be used. However, in the interplanetary trajectory, the gravitational force from other sources such as the Sun and the target planet or the Moon becomes prominent as a spacecraft moves far from the Earth. In the interplanetary trajectory, it is more convenient to use the SE coordinate system for defining the state space model of the spacecraft. In this case, the acceleration of the spacecraft can be expressed using (2.25) and (2.48) considering the Sun as the central body. The state space model becomes:

$$\begin{bmatrix} \dot{\boldsymbol{r}} \\ \dot{\boldsymbol{v}} \end{bmatrix} = \begin{bmatrix} \boldsymbol{v} \\ -\frac{GM_s}{r^3}\boldsymbol{r} + \sum_{j=1}^{n_p} GM_j \left(\frac{\boldsymbol{r}_{js}}{r_{js}^3} - \frac{\boldsymbol{r}_j}{r_j{}^3} \right) \end{bmatrix} \tag{2.63}$$

Here M_s is the mass of the Sun and vectors $\boldsymbol{r}$ and $\boldsymbol{v}$ are defined in the SE coordinate system.

References

[1] D. A. Vallado, *Fundamentals of Astrodynamics and Applications*. Springer Science & Business Media, 2001, vol. 12.

[2] R. R. Bate, D. D. Mueller, and J. E. White, *Fundamentals of Astrodynamics*. New York: Dover Publications, 1971.

[3] "National Geospatial-Intelligence Agency (Nga) Standardization Document," National Geospatial-Intelligence Agency (NGA), Tech. Rep., 2014.

[4] O. Montenbruck and E. Gill, *Satellite Orbits: Models, Methods, and Applications*, 1st ed. Springer-Verlag, 2002.

[5] J. R. Vetter, "The Evolution of Earth Gravitational Models Used in Astrodynamics,"

[6] B. D. Tapley, B. E. Schutz, and G. H. Born, *Statistical Orbit Determination*. Amsterdam ; Boston: Elsevier Academic Press, 2004.

[7] S. Pines, "Uniform Representation of the Gravitational Potential and its Derivatives," *AIAA Journal*, vol. 11, no. 11, pp. 1508–1511, Nov. 1973.

[8] J. B. Lundberg and B. E. Schutz, "Recursion formulas of Legendre functions for use with nonsingular geopotential models," *Journal of Guidance, Control, and Dynamics*, vol. 11, no. 1, pp. 31–38, Jan. 1988.

[9] D. A. Vallado and D. Finkleman, "A critical assessment of satellite drag and atmospheric density modeling," *Acta Astronautica*, vol. 95, pp. 141–165, 2014.

[10] E. M. Gaposchkin, "Calculation of satellite drag coefficients," Massachusetts Institute Of Technology Lincoln Laboratory, Tech. Rep., 1994.

[11] B. P. Graziano, "Computational modelling of aerodynamic disturbances on spacecraft within a concurrent engineering framework," Ph.D. dissertation, Cranfield University, 2007.

[12] "US Standard Atmosphere," NOAA, Tech. Rep., 1976.

[13] H. D. Curtis, *Orbital Mechanics for Engineering Students* (Elsevier Aerospace Engineering Series), 1. ed., reprinted. Amsterdam: Elsevier/Butterworth-Heinemann, 2008.

[14] S. K. Biswas, "Computationally Efficient Non-Linear Kalman Filters for On-Board Space Vehicle Navigation," Ph.D. dissertation, UNSW Sydney, Sydney, Australia, 2017.

Chapter 3

Observations

3.1 OVERVIEW

The space vehicle navigation or tracking process starts with a set of observations or measurements. Although we are interested in determining a spacecraft's position and velocity vectors, we cannot measure these vectors directly. Rather some scalar measurements are recorded, and a set of such scalar measurements are used to estimate the position and velocity of the spacecraft. For mathematical abstraction and brevity let us define the state vector $\boldsymbol{x}(t) = [\boldsymbol{r}(t)\ \boldsymbol{v}(t)]^T$ and $\boldsymbol{x}(t) \in \mathbb{R}^6$ and include m scalar measurements at an epoch in a single vector $\boldsymbol{z}(t) \in \mathbb{R}^m$. We can define the observation model as $\boldsymbol{h} : \mathbb{R}^6 \to \mathbb{R}^m$ and $\boldsymbol{z}(t) = \boldsymbol{h}(\boldsymbol{x}(t))$. At a later stage, we will see that the dimension of the state vector can be more than 6 depending on the estimation algorithm design. However, in this chapter, we will consider $\boldsymbol{x}(t) \in \mathbb{R}^6$ and develop observation models accordingly.

The observation model $\boldsymbol{h}$ can be linear or nonlinear or any function in a general sense. However, for space vehicle navigation and tracking, this function is nonlinear and stochastic. In this chapter, we will discuss various observation models for different measurement methods and strategies.

Ground-based navigation and tracking of space vehicles date back to the beginning of the space age. A ground-based tracking station, *Minitrack*, was established in the late 1950s in the United States to track Vanguard satellites [1]. This system measured a single set of angle observations. Generally, multiple scalar observations related to the position and velocity of spacecraft are stacked as a vector and used for navigation. Widely used observations are range, range rate, azimuth, and elevation angle of spacecraft with respect to the observation facility. Modern spacecraft navigation techniques can be segregated into two different strategies:

1. Ground-based observations
2. onboard observations

This chapter will introduce various observation techniques covering both the strategies mentioned above and discuss mathematical models of these observations. We will observe the use of these models in later chapters in the context of some specific applications.

3.2 GROUND-BASED OBSERVATIONS

In ground-based spacecraft navigation strategies, the measurements are performed from precisely known locations on the ground. The measurements are used in estimation algorithms to calculate the position and velocity of the satellite. Using high-fidelity dynamic models of satellite motion with ground-based observations can produce subdecimeter-level accuracy in position and sub-mm/s accuracy in velocity estimation [2]. The Deep Space Network (DSN), Satellite Laser Ranging (SLR) system, and DORIS system are the widely used ground-based satellite navigation systems. These systems can be classified as two-way ranging systems where the station on the ground transmits the signal, the target satellite acts as a transponder/reflector, and the ground station receives the signal from the satellite. In [3] the use of spaceborne GPS observations were recommended for on-ground orbit determination. Currently, GPS is also used as a part of ground-based navigation systems where the GPS flight data from the satellite is transmitted to the ground station, and the data is postprocessed to estimate the position and velocity of the satellite very precisely [4]. Figure 3.1 shows the basic functionality of ground-based navigation.

3.2.1 Active and Passive Radio Observations

Ground-based radar provides range and antenna pointing angles (i.e., azimuth and elevation) as observations. The relation of the radar observations with the space vehicle position can be conveniently expressed in the topocentric frame. In the topocentric ENU frame with the origin at the radar site, the position vector $\boldsymbol{r}_T$ of the space vehicle is defined as

$$\boldsymbol{r}_T = \begin{bmatrix} r_{T1} \\ r_{T2} \\ r_{T3} \end{bmatrix} = r \begin{bmatrix} \cos\beta\cos\theta \\ \cos\beta\sin\theta \\ \sin\beta \end{bmatrix} \tag{3.1}$$

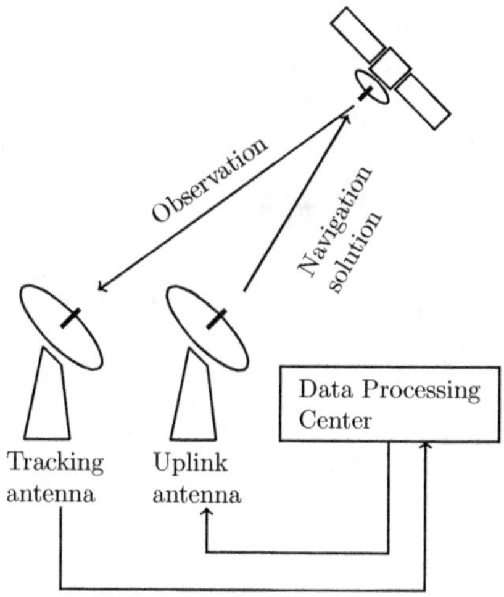

Figure 3.1: Ground-based satellite navigation.

where r is the true range of the space vehicle and θ and β are the azimuth and elevation, respectively. The one-way range is measured from the difference between the transmission and reception time of the signal. The observed range differs from the ideal or true geometric range of the space vehicle due to the clock biases of the transmitter and receiver and the transmission delay due to the atmosphere. Also, the measurement is corrupted by random noise. Mathematical realization of the range observation $\tilde{r}$ is

$$\tilde{r} = r + \delta r_b + \delta r_{atm} + \epsilon \tag{3.2}$$

where δr_b is the error due to the clock bias, δr_{atm} is the error due to the atmospheric delay, and ϵ is the random noise.

Similarly, considering the bias and noise, the measured azimuth $\tilde{\theta}$ and elevation $\tilde{\beta}$ angles can be modeled as

$$\tilde{\theta} = \tan^{-1} \frac{r_{T2}}{\sqrt{r_{T1}^2 + r_{T2}^2}} + b_\theta + \epsilon_\theta \tag{3.3}$$

and

$$\tilde{\beta} = \tan^{-1} \frac{r_{T3}}{\sqrt{r_{T1}^2 + r_{T2}^2}} + b_\beta + \epsilon_\beta \tag{3.4}$$

where, b_θ, b_β are biases in the azimuth and elevation measurements and ϵ_θ and ϵ_β are the measurement noises, respectively.

Considering $\boldsymbol{v}_T$ as the relative spacecraft velocity with respect to the ground station in the topocentric frame, one could model the range rate measurement as

$$\dot{\hat{r}} = \boldsymbol{r}_T^T \boldsymbol{v}_T + b_{\dot{r}} + \epsilon_{\dot{r}} \tag{3.5}$$

where $b_{\dot{r}}$ and $\epsilon_{\dot{r}}$ are bias and noise in range rate measurement.

3.2.2 Satellite Laser Ranging

SLR can be used for range measurement for satellites fitted with laser retroreflectors. Essentially the round-trip light time from the SLR ground station to the target satellite is measured. The range measurement ρ can be modeled as:

$$\rho = \frac{1}{2}\left(\rho_{up} + \rho_{down}\right) + \Delta\rho_{rel} + \Delta\rho_{atm} + \Delta\rho_R + \epsilon \tag{3.6}$$

Here, ρ_{up} and ρ_{down} are uplink and downlink distances between the ground station and the target satellite, $\Delta\rho_{rel}$ is the relativistic correction, $\Delta\rho_{atm}$ is the delay due to atmosphere, $\Delta\rho_R$ is the retroreflector specific correction, and ϵ is random noise, which can be assumed to be Gaussian.

3.2.3 Very Long Baseline Interferometry

Azimuth and elevation in the local tangent coordinate system of the ground station are obtained by measuring the direction of the maximum signal amplitude of the spacecraft. For very accurate angle measurement, very long baseline interferometry (VLBI) is used. In this technique, a spacecraft is tracked by two ground stations situated on different continents. The electromagnetic wavefront coming from the spacecraft reaches the two stations at different times. From time delay, the angular position is calculated as shown below.

In Figure 3.2 wavefront from a distant source arrives as a plane wave at two widely separated stations. The recorded signals are subsequently cross-correlated to determine the difference in the signal time of arrival at the two stations. This time difference is called VLBI delay and comprises geometric delay plus delay due to station clock offsets and differences in signal delays through the ionosphere, troposphere, instrumentation, and so forth. The geometric delay can be expressed as

$$\tau_g = \frac{1}{c}\mathbf{B} \cdot \hat{s} \tag{3.7}$$

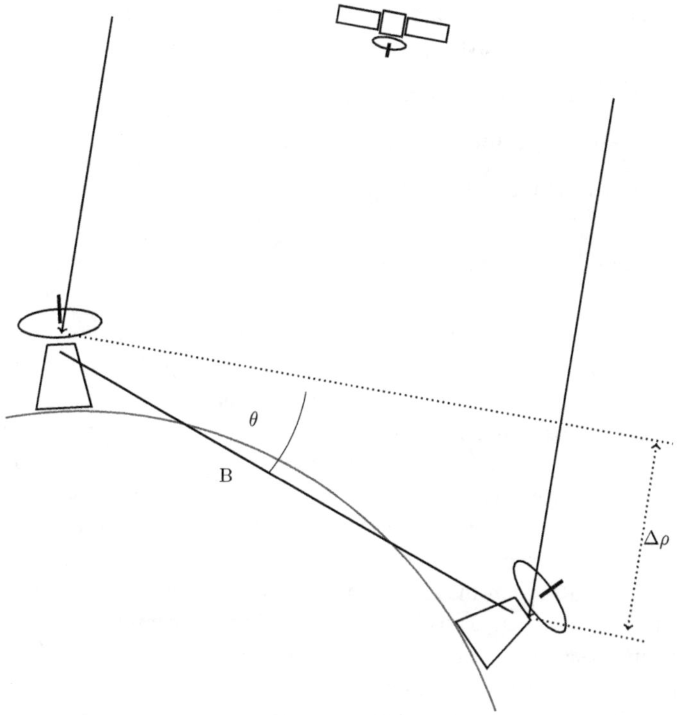

Figure 3.2: VLBI.

where **B** is the baseline vector between two ground stations and $\hat{s}$ is the source direction. So, from prior knowledge of the ground station positions, one angular component of the spacecraft can be calculated from the time delay. Errors can be reduced by measuring the VLBI delay of quasars and subtracting this from the VLBI delay of spacecraft. This technique is called the delta VLBI method.

3.3 ONBOARD OBSERVATIONS

The idea of autonomous spacecraft navigation was conceived in the 1970s to reduce onboard latency, cost, and processing time of received satellite data. Lowrie documented various potential autonomous navigation systems in [5]. Based on the measurement technique adopted, the autonomous navigation systems can be categorized into three groups [5]:

1. Angular measurements to celestial objects
2. Earth-based target reference measurement
3. Range measurement to known beacon

3.3.1 Angular Measurements to Celestial Objects

In this method, the angles to three different stars with respect to the local vertical to the Earth's surface are measured [5]. It is also known as celestial navigation and is a very old navigation technique used for maritime navigation. This method enables determining the position and attitude of a satellite without using any external support from either ground or space. Several configurations are used in this method, for example, the angle between Sun and local vertical, Moon and star, or planet and star. This is a well-established technique in the satellite navigation community and is known for inherent inaccuracy in position determination. The navigation solution is highly sensitive to angular measurements. Instruments required to measure precise and accurate angles are also costly. This results in higher mission cost.

The approach towards the navigation solution using this method also changed over time. In 1972, Srivastava proposed a navigation solution for Lincoln Experimental Satellites using angular measurements of the Sun. These measurements were used to determine the inclination of the satellite orbit. When the Kalman filter became a mainstream and well-proven estimation method, the solution interest shifted towards position estimation in a Cartesian reference frame. Recent work on celestial navigation using an unscented Kalman filter and particle filter shows that the real-time RMS position and velocity error can be reduced to 103 m and 0.16 m/s, respectively [6]. The result implies that celestial navigation cannot provide the desired accuracy for precise real-time position and velocity estimation. This system is used along with other sensors for improved accuracy for navigation in deep space [7].

3.3.2 Earth-Based Target Reference Measurement

This system relies on ground control points capable of emitting electromagnetic signals or detecting landmarks on the Earth. LANDSAT satellites, using multispectral scanners, were able to detect high-intensity Xenon lamps. Draper Labs suggested the use of similar artificial landmarks for navigational purposes [5]. But the detection of passive landmarks on the Earth is a more independent solution. Image processing is used to detect landmarks. The orientation of the landmark in the field of view of the satellite camera and the attitude information are used to calculate the position of the satellite. This technique has been extended to identifying landmarks on other celestial objects like planets, the Moon, and asteroids. Cheng et el. documented landmark detection-based autonomous methodology as being faster than manual landmark detection used in the Near-Earth Asteroid Rendezvous (NEAR) mission in [8], which can provide 100 m accuracy in position estimation. The conclusion was drawn based on simulation results; however, it was the first initiative to integrate this type of technology in the autonomous navigation of spacecraft [8]. Trawny et al. performed field experiments using helicopters and subsonic parachute drop tests to observe the capability of landmark-based systems for the autonomous landing of spacecraft [9]. Vision-based observations were integrated with an inertial measurement unit (IMU) using an EKF to estimate position, velocity, and attitude. This technology has been proven effective for the entry, descent, and landing phases of interplanetary spacecraft.

3.3.3 Range Measurement to Known Beacon

Range measurement to a known beacon provides a position solution using trilateration. GPS or, in a more general sense, GNSS works based on this principle. GPS was conceived in 1973 for the tracking of military assets. Lowrie suggested the use of this system for spaceborne activities [5]. Landsat-4 was the first satellite to carry a spaceborne GPS receiver [10]. At that time, GPS was not fully functional, and the navigation system was not autonomous. Landsat-4 relied on processing GPS measurement data on the ground. For precise positioning, by postprocessing, the GPS flight data has been a well-accepted technique since then. The success of the TOPEX/Poseidon, GRACE, and CHAMP missions has proven the use of GPS in LEO missions to be a low-cost and simple solution. Millimeter-level relative positioning accuracy can be achieved by using differential GPS measurements in space [11]. But the real-time absolute navigation solution accuracies using GPS are still 10 m and 0.1 m/s for position and velocity, respectively. Also, other positioning sensors and ground-based navigation are still in use in addition to GPS-based autonomous navigation because of the chances of blind spots and outliers in measurements. In recent years, new GNSS like Galileo, and Regional Navigation Satellite System (RNSS), for example, Navigation using Indian Constellation (NavIC), and Quasi-Zenith Satellite System (QZSS) have become fully operational. This has opened a new possibility of developing an autonomous satellite navigation system using multi-GNSS measurements, which is more reliable, highly accurate, and does not require any ground support for postprocessing. Development of this type of system can be cost-effective due to the lower dependency on ground stations.

3.3.4 GNSS Observations

GNSS satellites broadcast the transmission time and their corresponding position. A GNSS receiver acquires the signal and, using the internal clock, calculates the time of reception. The time difference between transmission and reception is utilized to calculate the range. But this range does not reflect the true range between the satellite and the receiver due to the errors. The sources of various errors are shown in Figure 3.3. Code and carrier phase measurements are the two available observations from GNSS [12]. The carrier phase measurement is more accurate than code phase but suffers from integer ambiguity. For a spaceborne receiver, both the observations are affected by [12]:

1. *Satellite clock error*: This error occurs due to the bias and drift of the GNSS satellite clock. The satellite clock bias and drift error are broadcast through the GNSS signal and utilized for the range correction. For precise positioning, the International GNSS Services (IGS) provides highly accurate satellite clock bias and drift estimation [13]. However, this support is not available in real-time.

2. *Receiver clock error*: Receiver clock error is the result of the receiver clock drift and bias. These parameters are estimated by including them in the state vector.

3. *Ephemerides prediction error*: The position information of a transmitting GNSS satellite is obtained by propagating the GNSS broadcast ephemerides. The ephemerides are not updated frequently, resulting in a position error when propagated for a longer time. For precise positioning, the IGS ephemeris product must be used to correct the error. This error can be reduced in real time by using an empirical propagation model [14].

4. *Atmospheric delay*: The atmosphere influences electromagnetic wave propagation and causes a delay in the signal reception. This delay can be further segregated to the tropospheric and the ionospheric delay. The tropospheric delay is calculated using the Saastamoinen model. Ionospheric delay can also be calculated using a worldwide ionospheric model [15].

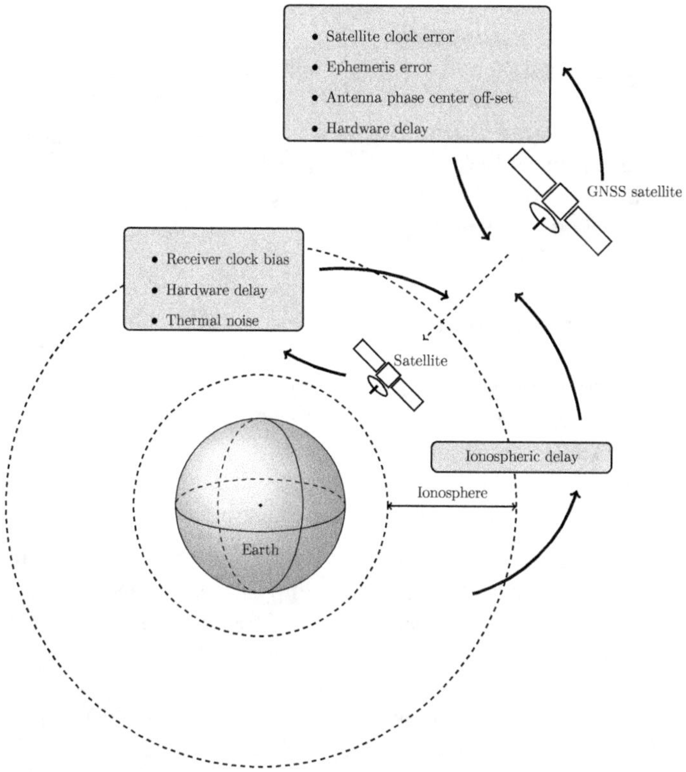

Figure 3.3: Errors in GNSS observation.

5. *Receiver noise*: This is the result of thermal noise in the receiver and is considered as random noise in the measurement model.

In a general sense GNSS range observations are one-way range observations. Range observations from code phase measurements are called pseudoranges and those from carrier phase measurements are called carrier ranges. Pseudorange and carrier range measurements of the GPS are modeled as [12]:

$$\rho_i(t) = r_i(t) + c[\delta t_u(t) - \delta t_i(t-\tau)] + I(t) + T(t) + \epsilon_\rho(t) \quad (3.8)$$

$$\Phi_i(t) = r_i(t) + c[\delta t_u(t) - \delta t_i(t-\tau)] + I_\phi(t) + T_\phi(t) + \lambda N + \epsilon_\Phi(t) \quad (3.9)$$

where

i	is the GNSS satellite index
ρ_i	is the pseudorange from the user vehicle to the navigation satellite i
Φ_i	is the carrier range from the user vehicle to the navigation satellite i
r_i	is the geometric distance from the user vehicle to the navigation satellite i
δt_u	is the receiver clock bias
δt_i	is the clock bias of the navigation satellite
τ	is the signal transmission time
c	is the velocity of light
$I(t)$	is the ionospheric error for pseudorange
$T(t)$	is the tropospheric error for pseudorange
$I_\Phi(t)$	is the ionospheric error for carrier range
$T_\Phi(t)$	is the tropospheric error for carrier range
λ	is the wavelength of the carrier signal
N	is the integer ambiguity
$\epsilon_\rho(t)$	is the random noise in pseudorange measurement
$\epsilon_\Phi(t)$	is the random noise in carrier range measurement

Receiver clock error is estimated by including this bias error in the state vectors [12]. For a 10 m level of accuracy in position, ephemeris prediction error is neglected [16] in onboard operation. But Montenbruck has shown that if a real-time precise ephemeris of GPS satellites is broadcast using the TDRSS, it is possible to estimate the position of a satellite carrying a GPS receiver much more precisely [2]. This proposition requires a dependency on the TDRSS system, and also, the level of dependency onground support will increase. A dual-frequency GPS receiver can be used to eliminate ionospheric error. Yunck suggested the group and phase ionospheric correction (GRAPHIC) technique, which uses code and carrier phase measurement to calculate ionospheric delay-free measurements [17]. This technique does not require a dual-frequency receiver and has been practiced in the navigational community for a long time [18]. However, the integer ambiguity must be resolved before using this method. For multi-GNSS applications, intersystem clock bias of multiple navigation constellations must be considered in the measurement model and must be estimated [19]. For example, if the GPS and Galileo constellations are used, then the receiver clock bias for Galileo is modeled as

$$\delta t_{GAL} = \delta t_{GPS} + \delta t_{ISB} \tag{3.10}$$

where δt_{GAL}, δt_{GPS}, and δt_{ISB} are the receiver clock bias for Galileo, the receiver clock bias for GPS, and the intersystem bias (ISB), respectively. The receiver clock bias is modeled as a first-order Markov process. The dynamics of the receiver clock bias vector can be represented as:

$$\begin{bmatrix} \dot{\delta t}_{GPS} \\ \dot{\delta t}_{ISB} \end{bmatrix} = \begin{bmatrix} d \\ 0 \end{bmatrix} + \begin{bmatrix} w_{GPS} \\ w_{ISB} \end{bmatrix} \tag{3.11}$$

where d is the receiver clock drift, and w_{GPS} and w_{ISB} are random noises. In this model, a constant receiver clock drift $\dot{\delta t}_{GPS}$ is assumed. Receiver noise is addressed by either introducing a weighting matrix in the least square estimation (LSE) or using a sequential estimation algorithm.

3.3.5 Dilution of Precision

The accuracy of the GNSS-based position solution depends on the geometry of the satellites in view. Let us analyze the position solution to understand the impact of the satellite geometry. For a user receiver position $\boldsymbol{r}_u = [x_u, y_u, z_u]^T$ and the i-th navigation satellite position $\boldsymbol{r}_i = [x_i, y_i, z_i]^T$, assuming no atmospheric delay and neglecting the satellite clock bias, the pseudorange measurement is given by

$$\begin{aligned} \rho_i &= \sqrt{(x_u - x_i)^2 + (y_u - y_i)^2 + (z_u - z_i)^2} + c\delta t_u + \epsilon \\ &:= f_i(x_u, y_u, z_u) + c\delta t_u + \epsilon \end{aligned} \tag{3.12}$$

Let the estimate of the user position be $\hat{r}_u = [\hat{x}_u, \hat{y}_u, \hat{z}_u]^T$ and the estimated receiver clock bias is $\hat{\delta t}_u$. The error in the estimate of the user position is given by

$$\begin{bmatrix} \Delta x_u \\ \Delta y_u \\ \Delta z_u \end{bmatrix} = \begin{bmatrix} x_u - \hat{x}_u \\ y_u - \hat{y}_u \\ z_u - \hat{z}_u \end{bmatrix} \tag{3.13}$$

Similarly, define the error in the estimate of the receiver clock bias as $\Delta t_u = \delta t_u - \hat{\delta t}_u$. Then the estimated pseudorange

$$\hat{\rho}_i = \sqrt{(\hat{x}_u - x_i)^2 + (\hat{y}_u - y_i)^2 + (\hat{z}_u - z_i)^2} + c\hat{\delta t}_u \tag{3.14}$$

We can express the measured pseudorange using the estimated user position as

$$\begin{aligned}
\rho_i &= f_i(x_u, y_u, z_u) + c\delta t_u + \epsilon \\
&= f_i(\hat{x}_u + \Delta x_u, \hat{y}_u + \Delta y_u, \hat{z}_u + \Delta z_u) + c(\hat{\delta t}_u + \Delta t_u) + \epsilon \\
&= f_i(\hat{x}_u, \hat{y}_u, \hat{z}_u) + \frac{\partial f_i}{\partial x_u}\Delta x_u + \frac{\partial f_i}{\partial y_u}\Delta y_u + \frac{\partial f_i}{\partial z_u}\Delta z_u + c(\hat{\delta t}_u + \Delta t_u) + \epsilon \\
&= \hat{\rho}_i + \frac{\partial f_i}{\partial x_u}\Delta x_u + \frac{\partial f_i}{\partial y_u}\Delta y_u + \frac{\partial f_i}{\partial z_u}\Delta z_u + c\Delta t_u + \epsilon
\end{aligned} \tag{3.15}$$

Subsequently, the pseudorange error is given by

$$\begin{aligned}
\Delta\rho_i &= \rho_i - \hat{\rho}_i \\
&= \frac{\partial f_i}{\partial x_u}\Delta x_u + \frac{\partial f_i}{\partial y_u}\Delta y_u + \frac{\partial f_i}{\partial z_u}\Delta z_u + c\Delta t_u + \epsilon
\end{aligned} \tag{3.16}$$

For a set of n satellites, the pseudorange error vector is given by

$$\begin{bmatrix} \Delta\rho_1 \\ \Delta\rho_2 \\ \vdots \\ \Delta\rho_n \end{bmatrix} = \begin{bmatrix} \frac{\partial f_1}{\partial x_u} & \frac{\partial f_1}{\partial y_u} & \frac{\partial f_1}{\partial z_u} & 1 \\ \frac{\partial f_2}{\partial x_u} & \frac{\partial f_2}{\partial y_u} & \frac{\partial f_2}{\partial z_u} & 1 \\ \vdots & \vdots & \vdots & \vdots \\ \frac{\partial f_n}{\partial x_u} & \frac{\partial f_n}{\partial y_u} & \frac{\partial f_n}{\partial z_u} & 1 \end{bmatrix} \begin{bmatrix} \Delta x_u \\ \Delta y_u \\ \Delta z_u \\ c\Delta t_u \end{bmatrix} + \begin{bmatrix} \epsilon_1 \\ \epsilon_2 \\ \vdots \\ \epsilon_n \end{bmatrix} \tag{3.17}$$

In compact form the above equation can be written as

$$\boldsymbol{\Delta}\rho = \boldsymbol{H}\boldsymbol{\Delta}x + \boldsymbol{\epsilon} \tag{3.18}$$

where $\boldsymbol{\Delta}\rho = [\Delta\rho_1, \Delta\rho_2, \ldots, \Delta\rho_n]^T$, $\boldsymbol{H}$ is the Jacobian matrix, and $\boldsymbol{\epsilon} = [\epsilon_1, \epsilon_2, \ldots, \epsilon_n]^T$. We will see in the next chapter that one can estimate $\boldsymbol{\Delta x}$ using the least squares method. The estimated user position error is then given by

$$E[\boldsymbol{\Delta x}] = (\boldsymbol{H}^T\boldsymbol{H})^{-1}\boldsymbol{H}^T\boldsymbol{\Delta}\rho \tag{3.19}$$

The error covariance

$$E[\boldsymbol{\Delta x}\boldsymbol{\Delta x}^T] = (\boldsymbol{H}^T\boldsymbol{H})^{-1}\boldsymbol{H}^T E[\boldsymbol{\Delta}\rho\boldsymbol{\Delta}\rho^T]\boldsymbol{H}(\boldsymbol{H}^T\boldsymbol{H})^{-1} \tag{3.20}$$

Considering equal standard deviation for all pseudorange observations, $\boldsymbol{E}[\boldsymbol{\rho}\boldsymbol{\rho}^T] = \sigma^2\boldsymbol{I}_{n\times n}$ and σ^2 is the variance of the pseudorange error. Then we can write

$$E[\boldsymbol{\Delta x}\boldsymbol{\Delta x}^T] = (\boldsymbol{H}^T\boldsymbol{H})^{-1}\sigma^2 \tag{3.21}$$

The previous equation signifies that the uncertainty in position and receiver clock bias estimation not only depends on the uncertainty in pseudorange measurements but also on the geometry of the satellites. The geometry of the satellites is represented by the Jacobian matrix $\boldsymbol{H}$. Let us define the estimation error covariance as

$$E[\boldsymbol{\Delta x \Delta x}^T] = \begin{bmatrix} \sigma^2_{x_u} & \sigma^2_{x_u y_u} & \sigma^2_{x_u z_u} & \sigma^2_{x_u c\delta t_u} \\ \sigma^2_{y_u x_u} & \sigma^2_{y_u} & \sigma^2_{y_u z_u} & \sigma^2_{y_u c\delta t_u} \\ \sigma^2_{z_u x_u} & \sigma^2_{z_u y_u} & \sigma^2_{z_u} & \sigma^2_{z_u c\delta t_u} \\ \sigma^2_{t_u x_u} & \sigma^2_{t_u y_u} & \sigma^2_{t_u z_u} & \sigma^2_{c\delta t_u} \end{bmatrix} \tag{3.22}$$

Then

$$\sigma^2_{x_u} + \sigma^2_{y_u} + \sigma^2_{z_u} + \sigma^2_{c\delta t_u} = tr((\boldsymbol{H}^T\boldsymbol{H})^{-1})\sigma^2 \tag{3.23}$$

The geometric dilution of precision (GDOP) is defined as

$$GDOP = \frac{\sqrt{\sigma^2_{x_u} + \sigma^2_{y_u} + \sigma^2_{z_u} + \sigma^2_{c\delta t_u}}}{\sigma} \tag{3.24}$$

Consequently

$$GDOP = \sqrt{tr((\boldsymbol{H}^T\boldsymbol{H})^{-1})} \tag{3.25}$$

Additionally, the position dilution of precision (PDOP) is defined as

$$PDOP = \frac{\sqrt{\sigma^2_{x_u} + \sigma^2_{y_u} + \sigma^2_{z_u}}}{\sigma} \tag{3.26}$$

Essentially $PDOP$ is the square root of the sum of the first three diagonal elements of the $(\boldsymbol{H}^T\boldsymbol{H})^{-1}$ matrix.

3.3.6 Inertial Navigation System

An INS consists of three accelerometers and three gyroscopes and does not require any external signal sources for operation. INS is used in launch, reentry, and landing missions in addition to other sensors. The accelerometers measure changes in the object's velocity, while the gyroscopes measure changes in the object's orientation. By integrating these measurements over time, an INS can determine the object's position and velocity relative to its starting point, as well as its orientation relative to a fixed reference frame. There are various types of gyroscopes and accelerometers are available. Gyroscopes can be mechanical, optical, or microelectromechanical systems (MEMS)-based. Similarly, mechanical, solid-state, and MEMS-based accelerometers are also available. However, MEMS-type gyroscopes and accelerometers are preferred in recent years due to their smaller number of moving parts, and low production and maintenance costs.

INS can be of two types — stable platform and strap-down. In the stable-platform INS system, the accelerometers and gyroscopes are mounted on a gimbaled platform. This platform is always kept aligned with a predefined global frame of reference using the gyroscope feedback to the motors of the gimbal. The gimbal angle is the measure of the orientation and the position and velocity are obtained by integrating the accelerometer outputs. The accelerometers and the gyroscopes in a strap-down system are tightly strapped on the vehicle itself. As a result, INS outputs are in the body frame of the vehicle. In the body frame, the axes are defined with respect to the longitudinal and lateral axes of the spacecraft. Strap-down systems are simpler and more rugged, while stable-platform systems offer higher accuracy and can operate over a longer period of time without requiring recalibration. However, with recent improvements in MEMS, very accurate strap-down systems are also available. As a result, strap-down INS are increasingly being used in space vehicles.

Modern gyroscopes in the strap-down INS systems are rate gyroscopes (i.e., provide angular rate as output). The gyroscope measurement suffers from bias and white noise. The bias term is not a constant, and it is susceptible to random flickering. As a result, this term is also modeled as a random variable. The gyroscope observation model can be written as:

$$\tilde{\omega} = \omega + b_\omega + \epsilon_\omega \tag{3.27}$$

where $\tilde{\omega}$ is the measured angular rate, ω is the true angular rate, b_ω is the bias, and ϵ_ω is a zero mean white noise.

Similar to the gyroscope observation, the accelerometer observation can be modeled as

$$\tilde{a} = a + b_a + \epsilon_a \tag{3.28}$$

where, $\tilde{a}$ is the measured acceleration, b_a is a bias which fluctuates over time, and ϵ_a is a zero mean white noise.

Integration of accelerometer measurement provides velocity and integration of the velocity results in position. Note that the calculated position and velocity will be with respect to the body frame. Integration of the gyroscope measurements from each body axis results in angular orientation changes in each axis, which are used to calculate the transformation matrix from the body frame to a global frame. This transformation matrix is used to convert the position from the body frame to the global frame. Notice that due to the presence of bias and noise in the measurements of both the gyroscope and the accelerometer the navigation solution incurs a large error over time. Hence the biases in measurements need to be estimated. In addition, there can be misalignment between the body frame and the strap-down system which results in additional fixed errors. The misalignment error is estimated through calibration beforehand.

3.4 SUMMARY

In this chapter, we have discussed various types of observations that are used in space vehicle navigation and tracking. Irrespective of the measurement strategy, most of the time observations are nonlinearly related to the position and the velocity vector of the space vehicle. In addition, the measurements include constant and fluctuating biases and noises. Consequently, retrieval of the position and velocity from these observations becomes a challenging task. In the next chapter, we will discuss a range of estimation algorithms that are used to estimate the position and velocity of space vehicles as well as the biases mentioned previously from inaccurate and noisy observations.

References

[1] B. D. Tapley, S. Bettadpur, M. Watkins, and C. Reigber, "The gravity recovery and climate experiment: Mission overview and early results," *Geophysical Research Letters*, vol. 31, no. 9, May 2004.

[2] O. Montenbruck, M. Markgraf, M. Garcia-Fernandez, and A. Helm, "GPS for Microsatellites – Status and Perspectives," *Small Satellites for Earth Observation*, R. Sandau, H.-P. Röser, and A. Valenzuela, Eds., Dordrecht: Springer Netherlands, 2008, pp. 165–174.

[3] E. Jochim, E. Gill, O. Montenbruck, and M. Kirschner, "GPS based onboard and onground orbit operations for small satellites," *Acta Astronautica*, vol. 39, no. 9-12, pp. 917–922, Nov. 1996.

[4] B.-S. Lee, J.-c. Yoon, Y. Hwang, and J. Kim, "Orbit determination system for the KOMPSAT-2 using GPS measurement data," *Acta Astronautica*, vol. 57, no. 9, pp. 747–753, Nov. 2005.

[5] J. Lowrie, "Autonomous navigation systems technology assessment," *17th Aerospace Sciences Meeting*, New Orleans,LA,U.S.A.: American Institute of Aeronautics and Astronautics, Jan. 1979.

[6] X. Ning and J. Fang, "Spacecraft autonomous navigation using unscented particle filter-based celestial/Doppler information fusion," *Measurement Science and Technology*, vol. 19, no. 9, p. 095 203, Sep. 2008.

[7] S. Muñoz, J. Christian, and E. G. Lightsey, "Development of an End to End Simulation Tool for Autonomous Cislunar Navigation," *AIAA Guidance, Navigation, and Control Conference*, Chicago, Illinois: American Institute of Aeronautics and Astronautics, Aug. 2009.

[8] Y. Cheng and J. K. Miller, "Autonomous landmark based spacecraft navigation system," *13th Annual AAS/AIAA Space Flight Mechanics Meeting*, 2003.

[9] N. Trawny, A. I. Mourikis, S. I. Roumeliotis, A. E. Johnson, and J. F. Montgomery, "Vision-aided inertial navigation for pin-point landing using observations of mapped landmarks," *Journal of Field Robotics*, vol. 24, no. 5, pp. 357–378, May 2007.

[10] O. Montenbruck, M. Garcia-Fernandez, A. Helm, and M. Markgraf, "Gps for microsatellites – status and perspectives," *In Proceedings of 6th IAA Symposium on Small Satellites for Earth Observation, April 23-26, Berlin*, 2007.

[11] S. Leung and O. Montenbruck, "Real-Time Navigation of Formation-Flying Spacecraft Using Global-Positioning-System Measurements," *Journal of Guidance, Control, and Dynamics*, vol. 28, no. 2, pp. 226–235, Mar. 2005.

[12] E. D. Kaplan and C. Hegarty, Eds., *Understanding GPS: Principles and Applications* (Artech House Mobile Communications Series), 2nd ed. Boston: Artech House, 2006.

[13] O. Montenbruck, A. Hauschild, and P. Steigenberger, "Differential Code Bias Estimation using Multi-GNSS Observations and Global Ionosphere Maps: Mulit-GNSS DCB Estimation," *Navigation*, vol. 61, no. 3, pp. 191–201, Sep. 2014.

[14] J. L. Garrison and B. E. Eichel, "An Extended Propagation Ephemeris for GNSS," *Navigation*, vol. 53, no. 3, pp. 167–179, Sep. 2006.

[15] M. Angel and F. Fernandez, "Advanced Ionospheric Modelling for GNSS Single Frequency Users," *2006 IEEE/ION Position, Location, And Navigation Symposium*, Coronado, CA: IEEE, 2006, pp. 110–120.

[16] E.-J. Choi, J.-C. Yoon, B.-S. Lee, S.-Y. Park, and K.-H. Choi, "Onboard orbit determination using GPS observations based on the unscented Kalman filter," *Advances in Space Research*, vol. 46, no. 11, pp. 1440–1450, Dec. 2010.

[17] T. P. Yunck, "Coping with the atmosphere and ionosphere in precise satellite and ground positioning," *Environmental Effects on Spacecraft Positioning and Trajectories*, ser. Geophysical Monograph Series, A. V. Jones, Ed., Wiley, 1993.

[18] O. Montenbruck, P. Swatschina, M. Markgraf, S. Santandrea, J. Naudet, and E. Tilmans, "Precision spacecraft navigation using a low-cost GPS receiver," *GPS Solutions*, vol. 16, no. 4, pp. 519–529, Oct. 2012.

[19] J. Paziewski and P. Wielgosz, "Accounting for Galileo extendashGPS inter-system biases in precise satellite positioning," *Journal of Geodesy*, vol. 89, no. 1, pp. 81–93, Jan. 2015.

Chapter 4

Estimation Algorithms

4.1 OVERVIEW

Estimation is the process of determining an approximation of desired quantities from incomplete or uncertain data. Navigation of a space vehicle can be formulated as an estimation problem, where we are interested in determining the position and velocity of the spacecraft from a set of observations or a sequence of observations. The observations can be the angle of arrival, time of arrival (a range), or time difference of arrival in this context. Any physical observations are not perfect; there are always some systematic and random errors. Depending on the mission, various types of estimation algorithms are used to determine the position and velocity of spacecraft as accurately as possible from noisy and erroneous observations.

Estimation can be performed in two ways: batch estimation and sequential estimation. In batch estimation, all the observations are processed to estimate the desired state vector. This technique requires the storage of all the observations and processing time increases as the number of observations increases [1]. Hence this method is not suitable for a real-time estimation solution. On the other hand, a sequential estimator uses observations collected at an epoch to estimate a state vector at that epoch.

The weighted least square estimation (LSE) and the Kalman filter are the most popular estimation algorithms for recovering the position and velocity information of a space vehicle from noisy and erroneous measurements. The weighted least square method has been known since the 18th century. It was developed by Gauss for planetary orbit corrections. Compared to that the sequential counterpart of the LSE, the Kalman filter is a relatively recent algorithm, developed 63 years ago. Nevertheless, this algorithm is almost ubiquitous due to its efficient and fast performance, which is particularly suitable for real-time onboard space vehicle navigation.

The Kalman filter is a statistical approach to optimal state estimation for linear systems and measurements with random noise [2]. The EKF was developed to apply the Kalman filter framework in nonlinear systems [3]. The application of the EKF spans almost all engineering disciplines.

The classical Kalman filter was designed to address the estimation problem for linear systems [2]. The NASA Ames Research Center applied this optimal estimation formula for estimating the position and the velocity of a space vehicle. As the dynamics of a space vehicle are nonlinear, the system was linearized using a first-order Taylor series approximation around an operating region to calculate the conditional error covariance and the Kalman gain, which is the observer gain selected to minimize the error covariance [3, 4]. This is known as the EKF and is now widely used for numerous nonlinear estimation problems. However, this approach leads to a suboptimal solution to the nonlinear estimation problem [5, 6].

Athans et al. proposed a second-order approximation technique to improve the estimation performance [6]. This second-order filter requires the calculation of both the Jacobian and Hessian of the nonlinear system under consideration and proved to be computationally expensive. Then Nørgaard et al. presented an approximate derivative calculation procedure using Stirling's interpolation formula to avoid analytical calculation of the Jacobian and Hessian for the second-order filter [7].

Julier and Uhlmann in their seminal work on nonlinear filters showed a new approach to predict the mean state vector and the error covariance using deterministic sampling [8–10]. This approach is known as the unscented transform (UT) and the filter which uses the UT in the prediction step is widely referred to as the unscented Kalman filter. The UKF ensures the accuracy of at least the second-order Taylor series approximation without Jacobian and Hessian calculation.

In this chapter, we present the space vehicle navigation problem as an estimation problem and discuss the EKF, UKF, and particle filter algorithms that can be used to estimate the position and velocity of a spacecraft.

4.2 SPACE VEHICLE NAVIGATION AS AN ESTIMATION PROBLEM

The aim of space vehicle navigation is to estimate the position and velocity of a space vehicle based on a certain set of observations and the dynamics of the space vehicle motion. The dynamics (i.e., the mathematical model of the space vehicle motion) are not exact and hence sre represented using a stochastic differential equation, rather than a deterministic ordinary differential equation. It should also be noted that the dynamics as well as the observations are nonlinear functions of the position and velocity of the space vehicle. Hence, the space vehicle navigation problem can be written as

$$\dot{\boldsymbol{X}}(t) = \boldsymbol{f}(t, \boldsymbol{X}(t)) + \boldsymbol{\nu}(t) \tag{4.1}$$

$$\boldsymbol{Z}(k) = \boldsymbol{h}(\boldsymbol{X}(k)) + \boldsymbol{\omega}(k) \tag{4.2}$$

where t denotes continuous time, k is the discrete equivalent of t, and $\boldsymbol{X}$ is a n-dimensional state vector that includes the position and velocity vector of the space vehicle to be estimated from the discrete measurement $\boldsymbol{Z}(k)$. $\boldsymbol{f}$ and $\boldsymbol{h}$ are nonlinear functions, and $\boldsymbol{\nu}(t)$ and $\boldsymbol{\omega}(k)$ are zero mean white process and measurement noises, respectively.

4.3 LEAST SQUARE ESTIMATION AND BATCH PROCESSING

Least square estimation is a popular method for estimation when the state vector and set of observations form an overdetermined system of equations. Linearization is done when the system of equations is nonlinear. Considering (4.2), one can write

$$\begin{aligned} \boldsymbol{Z}(k) &= \boldsymbol{h}(\boldsymbol{X}(k)) + \boldsymbol{\omega}(k) \\ &\approx \boldsymbol{h}(\widehat{\boldsymbol{X}}(k)) + \frac{\partial \boldsymbol{h}}{\partial \boldsymbol{X}} \boldsymbol{\Delta X}(k) + \boldsymbol{\omega}(k) \end{aligned} \tag{4.3}$$

Here, $\widehat{\boldsymbol{X}}(k)$ is the estimated state and $\boldsymbol{\Delta X}(k) = \boldsymbol{X}(k) - \widehat{\boldsymbol{X}}(k)$. Defining $\boldsymbol{\Delta Z}(k) = \boldsymbol{h}(\boldsymbol{X}(k)) - \boldsymbol{h}(\widehat{\boldsymbol{X}}(k))$ and $\frac{\partial \boldsymbol{h}}{\partial \boldsymbol{X}} = \boldsymbol{H}(k)$

$$\boldsymbol{\Delta Z}(k) = \boldsymbol{H}(k)\boldsymbol{\Delta X}(k) + \boldsymbol{\omega}(k) \tag{4.4}$$

With this linearization, the problem is approximated as a linear least square estimation problem where the aim is to find a $\boldsymbol{\Delta X}(k)$ that minimizes $\boldsymbol{\Delta Z}(k)^T$-$\boldsymbol{\Delta Z}(k)$; that is, the squared error in the measurement. Mathematically, the problem is expressed as

$$\widehat{\boldsymbol{\Delta X}}(k) = \arg \min_{\boldsymbol{\Delta X}(k)} [\boldsymbol{\Delta Z}(k)^T \boldsymbol{\Delta Z}(k)] \tag{4.5}$$

where, $\widehat{\Delta X}(k)$ is an estimate of $\Delta X(k)$. It can be easily shown that $\Delta Z(k)^T \Delta Z(k)$ is minimum when

$$\widehat{\Delta X}(k) = (H(k)^T H(k))^{-1} H(k)^T \Delta Z(k) \tag{4.6}$$

Subsequently, $\widehat{X}(k)$ is corrected using $\widehat{\Delta X}(k)$. This is done iteratively until $|\widehat{\Delta X}(k)|$ becomes less than a very small predefined value (generally less than 10^{-3}). This method of estimation is called iterated least square estimation. In the case where the observations have different variances, a weighted least square approach can be adopted by defining a diagonal matrix W such that

$$W_{ii} = \frac{1}{\sigma_i^2}, \quad i = 1, 2, 3, ..., n \tag{4.7}$$

Here, W_{ii} is the i-th diagonal element of W and σ_i is the standard deviation of the i-th observation. The weight matrix can be used to obtain the least square solution using the following equation:

$$\Delta X(k) = (H(k)^T W^{-1} H(k))^{-1} H(k)^T W^{-1} \Delta Z(k) \tag{4.8}$$

This can be applied to a batch of observations collected over a time interval, which is essentially batch processing. Consider a set of observations taken at discrete epochs as follows

$$\{Z(0), 0; Z(1), 1; Z(2), 2; ...; Z(n), n\}, \quad n \in \mathbb{N} \tag{4.9}$$

To estimate the state $X(0)$ one can construct a the following vectors and matrix

$$Z_b = [Z(0) Z(1) ... Z(n)]^T \tag{4.10}$$

$$H_b = \begin{bmatrix} \frac{\partial Z(0)}{\partial X(0)} \\ \frac{\partial Z(1)}{\partial X(0)} \\ \vdots \\ \frac{\partial Z(n)}{\partial X(0)} \end{bmatrix} \tag{4.11}$$

One can use (4.6) to estimate $X(0)$. Batch processing is often utilized for the postprocessing of a large number of observations. It should be noted that for processing observations of a large time interval the H_b matrix also becomes large and, as a result, it takes a substantial amount of time to process. One can devise a recursive least square estimator as well [11], however the complexity increases for dynamic state estimation using the said method. Additionally, this can be challenging aboard a space vehicle due to computation and storage constraints. For a real-time navigation solution batch processing is not very useful.

4.4 THE KALMAN FILTER

Rudolf E. Kalman was one of the primary developers of the celebrated eponymous filter that has found its applications across diverse disciplines. In this section, we will discuss the Kalman filter in the space vehicle navigation context. From the control theory perspective, the Kalman filter is an observer that minimizes the mean square error in the state estimate. The Kalman filter is a filter in the sense that it reduces the effect of model and observation uncertainties in the estimate. The original Kalman filter was developed for linear stochastic systems and subsequently extended to nonlinear systems. In this section, we will introduce the linear Kalman filter first and then the EKF.

4.4.1 The Kalman Filter for Discrete Linear Systems

Consider the following linear discrete system and measurement equations

$$\boldsymbol{X}(k+1) = \boldsymbol{\Phi}(k)\boldsymbol{X}(k) + \boldsymbol{\nu}(k) \tag{4.12}$$

$$\boldsymbol{Z}(k) = \boldsymbol{H}(k)\boldsymbol{X}(k) + \boldsymbol{\omega}(k) \tag{4.13}$$

Here, $\boldsymbol{\nu}(k)$ and $\boldsymbol{\omega}(k)$ are process noise and measurement noise vectors, respectively. Both the process and measurement noises are assumed to be zero mean white Gaussian noise. Additionally, let $\boldsymbol{\Delta X}(k)$ be the estimation error and define the following matrices:

$\boldsymbol{P}_{XX}(k) = E[\boldsymbol{\Delta X}(k)\boldsymbol{\Delta X}(k)^T]$ is the error covariance matrix
$\boldsymbol{Q}(k) = E[\boldsymbol{\nu}(k)\boldsymbol{\nu}(k)^T]$ is the process noise covariance matrix
$\boldsymbol{R}(k) = E[\boldsymbol{\omega}(k)\boldsymbol{\omega}(k)^T]$ is the measurement noise covariance matrix

The Kalman filter operation can be thought of as a two-step process — prediction and correction. Let us assume that at an epoch k, the estimated state and associated error covariance are $\widehat{\boldsymbol{X}}^+(k)$ and $\widehat{\boldsymbol{P}}_{XX}^+(k)$, respectively. In the prediction step, the state vector and the error covariance are propagated to the next epoch and the corresponding observation is calculated using the following set of equations:

$$\widehat{\boldsymbol{X}}^{-}(k+1) = \boldsymbol{\Phi}(k)\widehat{\boldsymbol{X}}^{+}(k) \tag{4.14}$$

$$\widehat{\boldsymbol{P}}_{XX}^{-}(k+1) = \boldsymbol{\Phi}(k)\widehat{\boldsymbol{P}}_{XX}^{+}(k)\boldsymbol{\Phi}(k)^{T} + \boldsymbol{Q}(k) \tag{4.15}$$

$$\widehat{\boldsymbol{Z}}(k+1) = \boldsymbol{H}(k+1)\widehat{\boldsymbol{X}}^{-}(k+1) \tag{4.16}$$

Note that $\widehat{(.)}^{-}$ denotes the a priori value of a quantity which is predicted using the system and measurement models, and $\widehat{(.)}^{+}$ is the a posteriori value of the same quantity with the correction from the observation. The correction step starts with computing the Kalman gain $\boldsymbol{K}$:

$$\boldsymbol{K}(k+1) = \widehat{\boldsymbol{P}}_{XX}^{-}(k+1)\boldsymbol{H}(k+1)^{T}(\boldsymbol{H}(k+1)\widehat{\boldsymbol{P}}_{XX}^{-}(k+1)\boldsymbol{H}(k+1)^{T} + \boldsymbol{R}(k+1))^{-1} \tag{4.17}$$

Setting the Kalman gain using the above equation ensures that the a posteriori mean squared error is minimized. The a posteriori state and error covariance are updated using the observation and the Kalman gain using the following equations

$$\widehat{\boldsymbol{X}}^{+}(k+1) = \widehat{\boldsymbol{X}}^{-}(k+1) + \boldsymbol{K}(k+1)\left[\boldsymbol{Z}(k+1) - \widehat{\boldsymbol{Z}}(k+1)\right] \tag{4.18}$$

$$\begin{aligned}\widehat{\boldsymbol{P}}_{XX}^{+}(k+1) &= (\boldsymbol{I} - \boldsymbol{K}(k+1)\boldsymbol{H}(k+1))\widehat{\boldsymbol{P}}_{XX}^{-}(k+1)(\boldsymbol{I} - \boldsymbol{K}(k+1)\boldsymbol{H}(k+1))^{T} \\ &\quad + \boldsymbol{K}(k+1)\boldsymbol{R}(k+1)\boldsymbol{K}(k+1)^{T}\end{aligned} \tag{4.19}$$

Note that by definition the error covariance matrix is positive definite. The covariance update using (4.19) is known as Joseph form where the error covariance update equation is in quadratic form. This ensures positive definiteness of the error covariance matrix. Another update form is also available, however, that does not ensure the positive definiteness of the error covariance.

4.4.2 Extended Kalman Filter

The extended Kalman filter was developed to address the nonlinearity in system and measurement models. We will consider (4.1) and (4.2) as the system and the measurement models in this discussion on the EKF. Similar to the linear Kalman filter, the EKF algorithm can be considered a two-step process: prediction and correction. Considering two consecutive measurement epochs as t and $t + \delta t$, the a priori mean state vector $\widehat{\boldsymbol{X}}^{-}(t+\delta t)$ can be written as

$$\begin{aligned}\widehat{\boldsymbol{X}}^{-}(t+\delta t) &= E[\boldsymbol{X}(t+\delta t)|\widehat{\boldsymbol{X}}^{+}(t)] \\ &= \widehat{\boldsymbol{X}}^{+}(t) + \int_{t}^{t+\delta t} \boldsymbol{f}(\tau, \boldsymbol{X}(\tau))d\tau \\ &= \boldsymbol{F}(t, \widehat{\boldsymbol{X}}^{+}(t)) \end{aligned} \tag{4.20}$$

If we consider the time interval δt in a manner that $t = k\delta t$ where k is a positive integer, then k and $k+1$ are the discrete equivalent of epoch t and $t+\delta t$, respectively. It should be noted that an analytical solution to the above integration is not often possible and numerical integration is performed. $\boldsymbol{P}_{XX}^{-}(k+1)$ is computed by linearization around $\widehat{\boldsymbol{X}}(k)$:

$$\boldsymbol{P}_{XX}^{-}(k+1) = \left.\frac{\partial \boldsymbol{F}}{\partial \boldsymbol{X}}\right|_{\widehat{\boldsymbol{X}}^{+}(k)} \boldsymbol{P}_{XX}^{+}(k) \left.\frac{\partial \boldsymbol{F}}{\partial \boldsymbol{X}}\right|_{\widehat{\boldsymbol{X}}^{+}(k)}^{T} + \boldsymbol{Q}(k) \tag{4.21}$$

here $\boldsymbol{P}_{XX}^{+}(k)$ is the a posteriori error covariance at the kth epoch, $\boldsymbol{Q}k$ is the process noise covariance matrix. Considering small interval δt, $\frac{\partial \boldsymbol{F}}{\partial \boldsymbol{X}}$ is

$$\left.\frac{\partial \boldsymbol{F}}{\partial \boldsymbol{X}}\right|_{\widehat{\boldsymbol{X}}^{+}(k)} = e^{\boldsymbol{\mathcal{J}}\delta t} \tag{4.22}$$

where $\boldsymbol{\mathcal{J}}$ is the Jacobian of $\boldsymbol{f}$ evaluated at $\boldsymbol{X} = \widehat{\boldsymbol{X}}^{+}(k)$. It is assumed that $\boldsymbol{\mathcal{J}}$ remains unchanged during δt interval. In this continuous-discrete setup, often the covariance propagation is done by solving the following continuous time differential equation:

$$\dot{\boldsymbol{P}}_{XX}(t) = \frac{\partial \boldsymbol{f}}{\partial \boldsymbol{X}}\boldsymbol{P}_{XX}(t) + \boldsymbol{P}_{XX}(t)\frac{\partial \boldsymbol{f}}{\partial \boldsymbol{X}}^{T} + \boldsymbol{Q}(t) \tag{4.23}$$

where $\boldsymbol{Q}(t)$ is the continuous process noise covariance matrix. The predicted measurement $\widehat{\boldsymbol{Z}}^{-}(k+1)$ is

$$\widehat{\boldsymbol{Z}}^{-}(k+1) = \boldsymbol{h}(\widehat{\boldsymbol{X}}^{-}(k+1)) \tag{4.24}$$

In the correction step, the Kalman gain $\boldsymbol{K}$ is calculated using (4.17), and the a posteriori mean state vector $\widehat{\boldsymbol{X}}^{+}(k+1)$ and the error covariance $\boldsymbol{P}_{XX}^{+}(k+1)$ at the $k+1$-th time step are computed using (4.18) and (4.19) respectively. Unlike the linear Kalman filter, $\boldsymbol{H}= \left.\frac{\partial \boldsymbol{h}}{\partial \boldsymbol{X}}\right|_{\widehat{\boldsymbol{X}}^{-}(k+1)}$.

4.4.3 Unscented Kalman Filter

As discussed in the previous section, the EKF propagates the nonlinear system dynamics using numerical integration for a nonlinear estimation problem. However, the covariance propagation is done through linear approximation. This can result in an error in the estimation for a highly nonlinear estimation problem, such as reentry vehicle or launch vehicle tracking. The time evolution of the error covariance of a nonlinear stochastic dynamic system can be expressed using the Fokker-Planck equation [12], which is a partial differential equation. The exact propagation of the error covariance would require integration of the Fokker-Planck Equation for the particular system, which is computationally intensive. Julier and Uhlmann proposed the unscented Kalman filter (UKF) which approximates the a priori error covariance more accurately than the EKF.

In the UKF framework, the UT is used to calculate the predicted mean state vector and the error covariance. In the UT, $2n+1$ weighted samples or sigma points are calculated at the kth time step from the estimated mean state vector $\widehat{\boldsymbol{X}}^+(k)$ and the error covariance $\boldsymbol{P}_{XX}^+(k)$ [9]. The i-th sigma point $\boldsymbol{X}_i$ and the corresponding weight W_i are [9]

$$\boldsymbol{X}_i(k) = \begin{cases} \widehat{\boldsymbol{X}}^+(k) & ,(i=0) \\ \widehat{\boldsymbol{X}}^+(k) + \Delta\boldsymbol{X}_i & ,(i=1,2,3...2n) \end{cases} \tag{4.25}$$

$$W_i = \begin{cases} \frac{\kappa}{n+\kappa} & ,(i=0) \\ \frac{1}{2(n+\kappa)} & ,(i=1,2,3...2n) \end{cases} \tag{4.26}$$

and

$$\Delta\boldsymbol{X}_i = (\sqrt{(n+\kappa)\boldsymbol{P}_{XX}^+(k)})_i \quad for\ i = 1,2,3....n \tag{4.27}$$

$$\Delta\boldsymbol{X}_i = -(\sqrt{(n+\kappa)\boldsymbol{P}_{XX}^+(k)})_i \quad for\ i = n+1,2,3....2n \tag{4.28}$$

where $(\sqrt{(n+\kappa)\boldsymbol{P}_{XX}^+(k)})_i$ is the i-th column of the matrix $\sqrt{(n+\kappa)\boldsymbol{P}_{XX}^+(k)}$. Note that, computation of a square root of a matrix is required at this step. Any efficient and numerically stable matrix square root computation method can be used [9]. Since $\boldsymbol{P}_{XX}^+(k)$ is positive definite, one can use Cholesky decomposition to find the square root of the matrix. κ is a parameter that can be used for fine tuning and if the noise is considered Gaussian, then heuristically κ can be selected in such a way that $n+\kappa = 3$ [8]. The computed error covariance can lose the positive definiteness property for $\kappa < 0$ [13]. In this case, scaled UT can be used [14]. In the scaled UT, the sigma points are scaled by the following equation

$$\boldsymbol{X}_i'(k) = \boldsymbol{X}_0(k) + \alpha(\boldsymbol{X}_i(k) - \boldsymbol{X}_0(k)) \tag{4.29}$$

where, α is a positive scaling parameter and $0 < \alpha < 1$. These sigma points are propagated to the $(k+1)$th time step using (4.20). The predicted mean $\widehat{\boldsymbol{X}}^-$ at the $(k+1)$th time step can be computed as [9]

$$\begin{aligned}\widehat{\boldsymbol{X}}^-(k+1) &= \sum_{i=0}^{2n} W_i \boldsymbol{X}_i(k+1) \\ &= \sum_{i=0}^{2n} W_i \boldsymbol{F}(k, \boldsymbol{X}_i(k))\end{aligned} \tag{4.30}$$

and the predicted covariance [9]

$$\boldsymbol{P}_{XX}^-(k+1) = \sum_{i=0}^{2n} W_i \left[\boldsymbol{X}_i(k+1) - \widehat{\boldsymbol{X}}^-(k+1)\right]\left[\boldsymbol{X}_i(k+1) - \widehat{\boldsymbol{X}}^-(k+1)\right]^T \tag{4.31}$$

The predicted measurement is calculated from the computed measurements corresponding to the i-th propagated sigma point and can be expressed as [9]

$$\boldsymbol{Z}_i(k+1) = \boldsymbol{h}(\boldsymbol{X}_i(k+1)) \tag{4.32}$$

Then the predicted measurement $\widehat{\boldsymbol{Z}}^-$ at $(k+1)$th time step is [9]

$$\widehat{\boldsymbol{Z}}^-(k+1) = \sum_{i=0}^{2n} W_i \boldsymbol{Z}_i(k+1) \tag{4.33}$$

The innovation covariance of the predicted measurement, is [9]

$$\boldsymbol{P}_{ZZ}(k+1) = \sum_{i=0}^{2n} W_i \left[\boldsymbol{Z}_i(k+1) - \widehat{\boldsymbol{Z}}^-(k+1)\right]\left[\boldsymbol{Z}_i(k+1) - \widehat{\boldsymbol{Z}}^-(k+1)\right]^T + \boldsymbol{R} \tag{4.34}$$

The cross-covariance matrix is calculated by [9]

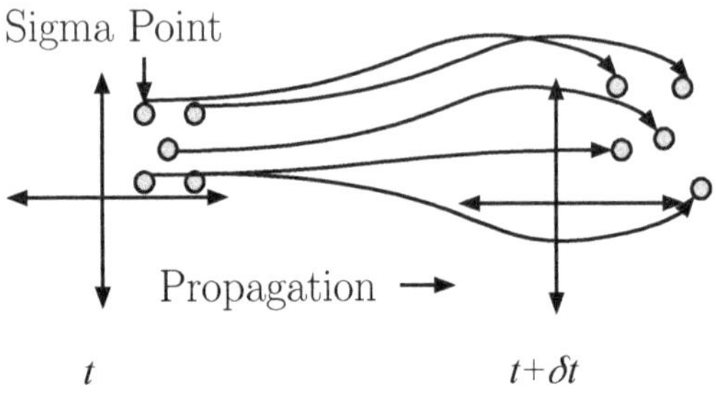

Figure 4.1: Propagation of sigma points using conventional UT.

$$P_{XZ}(k+1) = \sum_{i=0}^{2n} W_i \left[X_i(k+1) - \widehat{X}^-(k+1) \right] \left[Z_i(k+1) - \widehat{Z}^-(k+1) \right]^T \tag{4.35}$$

The correction step of the estimation can be performed by the standard Kalman filter equations

$$K(k+1) = P_{XZ}(k+1) P_{ZZ}(k+1)^{-1} \tag{4.36}$$

$$P^+_{XX}(k+1) = P^-_{XX}(k+1) - K(k+1) P_{ZZ}(k+1) K^T \tag{4.37}$$

$$\widehat{X}^+(k+1) = \widehat{X}^-(k+1) + K(k+1)[Z(k+1) - \widehat{Z}^-(k+1)] \tag{4.38}$$

It should be noted that (4.37) is not in quadratic form (i.e., this update does not ensure positive definiteness of the error covariance matrix). Here (4.37) is used because the Joseph form can be used only when the measurement function is linearized, which is the case for the EKF. This step may result in an ill-conditioned error covariance matrix or an error covariance matrix with very small negative eigenvalue(s) due to truncation and round-off errors in previous steps. This may result in numerical instability in the next prediction step during the computation of the square root of the error covariance matrix. A generalized Joseph formulation proposed by Zanetti and DeMars can be used in the UKF to address this issue [15].

Having discussed the UKF algorithm, let us now consider the computational aspect of the same. To determine the a priori mean and error covariance using the UT, $2n + 1$ sigma points must be propagated using (4.20). The UT approach is shown graphically in Figure 4.1. Most of the time a numerical integration technique is adopted for the state propagation using a differential equation, and in this case, the computation time for the UT will be much higher than that for the prediction steps in the EKF and KF. Let us consider a case where fourth-order Runge-Kutta numerical integration is adopted for the state propagation. Let $\widehat{\boldsymbol{X}}^{+}(t)$ be the a posteriori estimate of the state and $\boldsymbol{P}_{XX}^{+}$ be the estimated error covariance at time t. Using the Runge-Kutta technique at time $t+\delta t$ the first propagated sigma point will be,

$$\boldsymbol{X}_0(t+\delta t) = \widehat{\boldsymbol{X}}^{+}(t) + \frac{\delta t}{6}(\boldsymbol{k}_1 + 2\boldsymbol{k}_2 + 2\boldsymbol{k}_3 + \boldsymbol{k}_4) \tag{4.39}$$

where

$$\begin{aligned}
\boldsymbol{k}_1 &= \boldsymbol{f}(t, \widehat{\boldsymbol{X}}^{+}(t)) \\
\boldsymbol{k}_2 &= \boldsymbol{f}(t+\frac{\delta t}{2}, \widehat{\boldsymbol{X}}^{+}(t) + \frac{1}{2}\boldsymbol{k}_1\delta t) \\
\boldsymbol{k}_3 &= \boldsymbol{f}(t+\frac{\delta t}{2}, \widehat{\boldsymbol{X}}^{+}(t) + \frac{1}{2}\boldsymbol{k}_2\delta t) \\
\boldsymbol{k}_4 &= \boldsymbol{f}(t+\delta t, \widehat{\boldsymbol{X}}^{+}(t) + \boldsymbol{k}_3\delta t)
\end{aligned}$$

The above-mentioned operations must be performed $2n+1$ times to propagate all the sigma points. As a result, in every prediction step the function $\boldsymbol{f}$ must be evaluated $4(2n+1)$ times. However, for accuracy, if $\frac{\delta t}{h}$ $(h \in \mathbb{N})$ step size is chosen to propagate the state vector from time t to $t+\delta t$, then the function $\boldsymbol{f}$ must be evaluated $4h(2n+1)$ times. To account for process noise in the UKF, the state vector is augmented with process noise terms [8],

$$\boldsymbol{X}_a(t) = \begin{bmatrix} \boldsymbol{X}(t) \\ \boldsymbol{\nu}(t) \end{bmatrix} \tag{4.40}$$

and the sigma points for the UT are calculated from [8],

$$\widehat{\boldsymbol{X}}_a(t) = \begin{bmatrix} \widehat{\boldsymbol{X}}^{+}(t) \\ \boldsymbol{0}_{q\times 1} \end{bmatrix} \tag{4.41}$$

and

$$\boldsymbol{P}_a(t) = \begin{bmatrix} \boldsymbol{P}(t) & \boldsymbol{P}_{X\nu}(t) \\ \boldsymbol{P}_{X\nu}(t) & \boldsymbol{Q}(t) \end{bmatrix} \tag{4.42}$$

where $\boldsymbol{P}_{X\nu}$ is the cross covariance matrix of $\boldsymbol{X}(t)$ and $\boldsymbol{\nu}(t)$, q is the number of elements in $\boldsymbol{\nu}(t)$. The dimension of the augmented state vector is $m = n+q$. Then the number of sigma points to be propagated will increase to $2m+1$. This puts a substantial computational burden on the processor. Computation time is one of the major constraints of any real-time system. Also, for many real-time applications, computation power is limited due to several constraints. This makes the UKF difficult to implement in many real-time applications and the EKF remains the first choice of engineers.

To reduce the computation time of a UKF, the number of sigma points can be decreased. Various approaches have been proposed to improve the efficiency of the UKF, including the spherically simplex unscented transform (SSUT) which requires $n+2$ sigma points, $n+1$ of which lie on a hypersphere [16]. However, this can result in degraded estimation performance and only provides less than 50% in computation time compared to the UT [17]. The marginal unscented transformation (MUT) was suggested to reduce the number of sigma points for nonlinear functions with linear substructures [17]. For a problem with n state variables, if n_a are mapped nonlinearly, the number of sigma points can be reduced to $2n_a + 1$.

Recently the single propagation unscented Kalman filter (SPUKF) and the extrapolated single propagation unscented Kalman filter (ESPUKF) were suggested to significantly reduce the computation time of the UKF. In the subsequent part of this section, we will discuss these two methods.

4.4.4 Computationally Efficient Unscented Kalman Filters

The key idea in both the SPUKF and the ESPUKF is the approximation of all the sigma points at the current time step by propagating only one sigma point from the previous step. The a posteriori state vector of the previous time step is propagated to the current time step as per the EKF and then the other $2n$ (or $2m$) sigma points at the current time step are approximated from the deviations of the corresponding sigma points from the a posteriori state vector at the previous time step. Figure 4.2 shows the approach used in the SPUKF. The sigma points at time t are

$$\boldsymbol{X}_i(t) = \widehat{\boldsymbol{X}}^+(t) + \Delta\boldsymbol{X}_i, (i = 1, 2, 3, ...2n)$$

The i-th ($i \neq 0$) sigma point using the UT at time $t + \delta t$ will be

$$\begin{aligned}\boldsymbol{X}_i(t+\delta t) &= \boldsymbol{F}(t, \boldsymbol{X}_i(t)) \\ &= \boldsymbol{F}(t, \widehat{\boldsymbol{X}}^+(t)) + \boldsymbol{D}_{\Delta\boldsymbol{X}_i}\boldsymbol{F} + \frac{\boldsymbol{D}^2_{\Delta\boldsymbol{X}_i}\boldsymbol{F}}{2!} + \frac{\boldsymbol{D}^3_{\Delta\boldsymbol{X}_i}\boldsymbol{F}}{3!} + ... \end{aligned} \tag{4.43}$$

where

$$\boldsymbol{D}^p_{\Delta\boldsymbol{X}_i}\boldsymbol{F} = \left(\sum_{j=1}^{2n} \Delta X_{i_j} \frac{\partial}{\partial X_{i_j}}\right)^p \boldsymbol{F}(t, \boldsymbol{X}_i(t)) \tag{4.44}$$

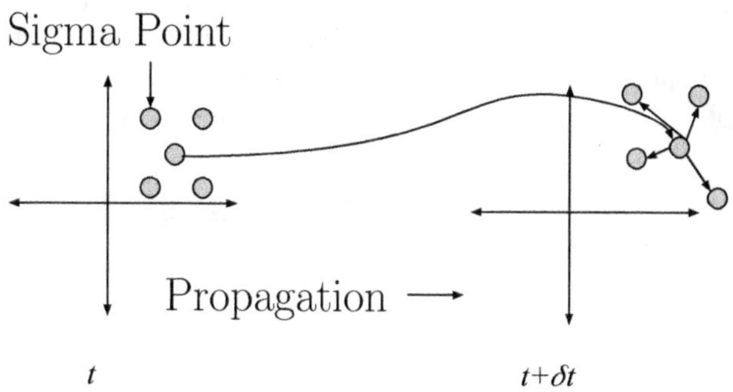

Figure 4.2: Approximate sigma point propagation.

and, ΔX_{i_j} and X_{i_j} are j-th component of ΔX_{i_j} and X_{i_j}, respectively. Considering first-order approximation,

$$\begin{aligned} \boldsymbol{X}_i(t+\delta t) &\approx \boldsymbol{F}(t, \widehat{\boldsymbol{X}}^+(t)) + \boldsymbol{D}_{\Delta \boldsymbol{X}_i}\boldsymbol{F} \\ &= \boldsymbol{X}_0(t+\delta t) + \left.\frac{\partial \boldsymbol{F}}{\partial \boldsymbol{X}}\right|_{\widehat{\boldsymbol{X}}^+(t)} \Delta \boldsymbol{X}_i \end{aligned} \tag{4.45}$$

Equation (4.45) creates the difference in the EKF approach and the SPUKF approach. In this context, note that the a priori state calculation is completely based on the propagation of the a posteriori mean state vector at the previous time step, whereas in the SPUKF the sigma points are approximated for a priori state vector computation.

To approximate all the sigma points at the current time, $\left.\frac{\partial \boldsymbol{F}}{\partial \boldsymbol{X}}\right|_{\widehat{\boldsymbol{X}}^+(t)}$ is evaluated once per step. This method reduces the number of times the function $\boldsymbol{f}$ needs to be evaluated to $4h$ (where h is the number of integration steps), instead of the original $4h(2n+1)$ times. Additionally, the Jacobian matrix and matrix exponential are evaluated once per step. This results in a significant decrease in processing time due to the reduced number of evaluations of the function $\boldsymbol{f}$. The steps for predicting measurements, computing the Kalman gain, calculating the mean state vector, and determining the error covariance are the same as those for the UKF [18].

Note that the SPUKF prediction step is accurate to the first-order Taylor series term. The ESPUKF utilizes the Richardson extrapolation method to include the second-order Taylor series terms in the sigma point propagation. This approach avoids the calculation of Hessian, which is difficult to evaluate for a complex multivariable nonlinear system. In the ESPUKF the sigma points are approximated after propagation of the mean state vector using the following equation:

$$\boldsymbol{X}_i(t+\delta t) = 2N_2(\frac{\boldsymbol{\Delta X_i}}{2}) - N_1(\boldsymbol{\Delta X_i}) \tag{4.46}$$

where

$$N_1(\boldsymbol{\Delta X_i}) = \boldsymbol{F}(t, \widehat{\boldsymbol{X}}^+(t)) + \boldsymbol{D}_{\boldsymbol{\Delta X}_i}\boldsymbol{F}|_{\widehat{\boldsymbol{X}}^+(t)} \tag{4.47}$$

and

$$N_2(\frac{\boldsymbol{\Delta X_i}}{2}) = \boldsymbol{F}(t, \widehat{\boldsymbol{X}}^+(t)) + \boldsymbol{D}_{\boldsymbol{\Delta X}/2}\boldsymbol{F}|_{\widehat{\boldsymbol{X}}^+(t)} + \boldsymbol{D}_{\boldsymbol{\Delta X}/2}\boldsymbol{F}|_{\widehat{\boldsymbol{X}}^+(t)+\frac{\boldsymbol{\Delta X_i}}{2}} \tag{4.48}$$

Considering $\boldsymbol{F}(t, \widehat{\boldsymbol{X}}^+(t)) \approx \boldsymbol{F}(t, \widehat{\boldsymbol{X}}^+(t) + \frac{\boldsymbol{\Delta X_i}}{2})$, it can be shown that the approximation error using (4.46) is of the order of the third-order Taylor series terms. It can be proved that the error in the state and the error covariance estimation using the ESPUKF is of the order of the third-order Taylor series terms [19]. As a consequence, the estimation accuracy of the ESPUKF is similar to that of the UKF. This method of sigma point propagation is termed as the extrapolated single propagation technique (ESPT).

The computation time improvement of the SPUKF and the ESPUKF depends on the dimension of the state vector n, number of the integration step size h and the computation complexity of the system dynamics j. The computation time improvement C_{SP} for the SPUKF is given by [18]

$$\begin{aligned} C_{SP} &= \frac{t_{UKF} - t_{SP}}{t_{UKF}} \times 100\% \\ &= \frac{(26h-4)n^2 + (8hj + 10h - j - 3)n - 5n^3}{26hn^2 + (8hj + 23h)n + 4hj + 5h} \times 100\% \end{aligned} \tag{4.49}$$

where t_{UKF} is the propagation time required per measurement update in the UKF and t_{SP} is the same for the SPUKF. Similarly, considering the propagation time per measurement update for the ESPUKF as t_{ESP}, the processing time improvement

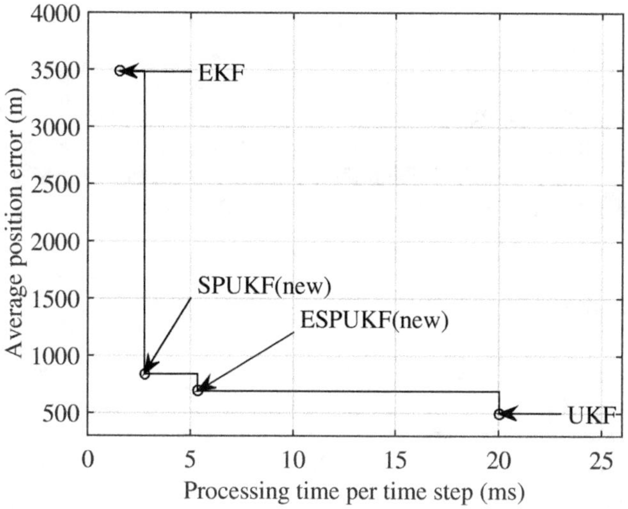

Figure 4.3: Processing time vs. estimation error.

for the ESPUKF is [18]

$$\begin{aligned} C_{ESP} &= \frac{t_{UKF} - t_{ESP}}{t_{UKF}} \times 100\% \\ &= \frac{(26h - 2k - 18)n^2 + (8hk + 10h - 3k - 4)n - 6n^4 - 11n^3}{26hn^2 + (8hj + 23h)n + 4hj + 5h} \times 100\% \end{aligned} \tag{4.50}$$

Equations (4.49) and (4.50) signify that the processing time of both the SPUKF or the ESPUKF will not improve for all estimation problems, and there can be cases where there will be no processing time improvement [18].

Figure 4.3 shows the processing time improvement of the SPUKF and ESPUKF in a reentry vehicle position estimation problem. In this problem, the altitude, down-range distance, and velocity of a reentry vehicle need to be estimated using range observations from a fixed known location. Using the reentry vehicle dynamics described in Chapter 2 and the radar range model mentioned in Chapter 3 various Kalman filters were implemented and the processing time per step was recorded.

It can be observed from Figure 4.3 that the average estimation error of the EKF for this problem is very large compared to that of the UKF, however, the processing time is very low. On the other hand, the UKF provides very accurate estimation performance at the cost of higher processing time. There is a modest improvement in the accuracy of the SPUKF compared to the EKF. A significant improvement is observed in terms of estimation accuracy for the ESPUKF and the performance is almost comparable to the UKF. For the ESPUKF approximately 75% processing time reduction compared to the UKF can be observed. It can be shown that the system complexity and the choice of integration step size affect the processing time reduction of the SPUKF and the ESPUKF [19].

4.5 PARTICLE FILTER

The discussion on state estimation for dynamic systems would be incomplete without introducing the readers to the particle filter, also known as Sequential Monte Carlo (SMC) method [20]. Particle filter-based estimation algorithms have been applied in diverse disciplines, for example, robotics [21], computer vision [22], and target tracking [23]. There are several examples of space application of the particle filter as well. Cheng et al. demonstrated the use of the particle filter for space vehicle attitude estimation [24]. Mashiku et al. used the particle filter in the orbit determination problem [25].

The core idea of the particle filter is to represent the probability distribution of the system state using a large set of weighted samples, which are referred to as particles. A large number of samples ensures an equivalent representation of the probability density function. Each particle represents a possible state of the system, and the weight of the particle reflects the likelihood of that state, given the available measurements. Mathematically the aposteriori PDF of the state vector $\boldsymbol{X}(k)$ given the observations till the k-th instance $\boldsymbol{Z}_{1:k}$ is approximated as

$$p(\boldsymbol{X}(k)|\boldsymbol{Z}_{1:k}) \approx \sum_{i=1}^{N_s} w_k^i \delta(\boldsymbol{X}(k) - \boldsymbol{X}^i(k)) \tag{4.51}$$

where N_s is the number of samples, w_k^i is the weight associated with the i-th sample at k-th epoch, and δ is the Dirac delta function.

The algorithm proceeds by first initializing the particle set with a prior distribution of the system state. Initially, all samples are assigned equal weights. Then, at each time step, the particles are propagated forward in time using a dynamic model that describes the evolution of the system state. Next, the weights of the particles are updated based on the current measurements. Essentially the output of the particle filter after the measurement update is an approximate PDF that is not necessarily Gaussian. One can take a weighted average of the particles to obtain the a posteriori mean state vector and subsequently compute the error covariance using the updated weights.

4.5.1 Sequential Importance Sampling

The weight update method in the particle filter is known as sequential importance sampling (SIS). In the SIS method, considering the system and measurement process as Markov processes, the updated weights at the k-th can be written as [20]

$$w_k^i \propto w_{k-1}^i \frac{p(\boldsymbol{Z}(k)|\boldsymbol{X}^i(k))p(\boldsymbol{X}^i(k)|\boldsymbol{X}^i(k-1))}{q(\boldsymbol{X}^i(k)|\boldsymbol{X}^i(k-1), \boldsymbol{Z}(k))} \tag{4.52}$$

Here, $q(.)$ is called the importance density from which the samples $\boldsymbol{X}^i, i = 1, 2, ..., N_s$ are drawn. Note that the updated weight is proportional to the likelihood of the particle given the measurement. After computing the weights for all the particles the weights are then normalized to preserve the property of the PDF.

One issue with importance sampling is that it can lead to the problem of weight degeneracy, where a few particles with high weights dominate the estimation, and the rest of the particles have negligible weights. This can be problematic because the estimation, in this case, relies only on a few particles that may not represent the PDF appropriately. To prevent degeneracy sequential importance resampling (SIR) is used where the particles are assigned new weights once the degeneracy is detected. The degeneracy in the particle filter is measured using the effective sample size N_{eff}, which is estimated using the following equation [20]

$$\widehat{N}_{eff} = \frac{1}{\sum_{i=1}^{N_s} (w_k^i)^2} \tag{4.53}$$

Essentially the estimated effective sample size $\widehat{N}_{eff} \leq N_s$ and very small value of $\widehat{N}_{eff}$ indicates severe degeneracy.

4.5.2 Particle Filter Variants

The particle filter has been utilized in various applications in real time, but its implementation has always been limited by the computation time required. The computational time and resource requirement for the particle filter arises due to a large number of particle propagation. Furthermore, as discussed earlier the particle filter has a problem called degeneracy. To address this problem, the particle flow filter was proposed by Daum et al., which uses log-homotopy and a partial differential equation (PDE) instead of importance sampling [26]. Later, Li et al. proposed a variant of the particle flow filter that efficiently calculates the weights of the particles using invertible mapping [27].

To overcome the computational challenges of the particle filter, several contributions have been made to develop computationally efficient versions. Sankaranarayanan et al. proposed algorithmic and architectural optimization for a particle filter for discrete systems and observations, which reduces processing time by using the independent metropolis hasting algorithm (IMHA) and auxiliary particle filtering [28]. Yi et al. suggested a computationally efficient joint sampling method for multitarget tracking using a particle filter through a sampling method called subsidiary importance sampling [29]. Berntorp et al. proposed an approximation of the measurement likelihood function using the Student-t distribution, which reduces the computation time of the particle filter for a specific class of discrete estimation problems [30].

In a recent development, the ESPT was applied to the particle filter framework for propagating particles [31]. This variant of the particle filter is referred to as the extrapolated single propagation particle filter (ESP-PF). This method demonstrated a significant processing time improvement compared to the classical particle filter due to the requirement of only one state vector propagation. Similar to the SPUKF and the ESPUKF, the computation time improvement of the ESP-PF depends on the complexity of the system dynamics, the number of integration steps and the number of particles to be propagated. The processing time improvement C_{ESP} is given by [31]

$$\begin{aligned} C_{ESP} &= \frac{t_{PF} - t_{ESP}}{t_{PF}} \times 100\% \\ &= \frac{(2-4N_s)n^2 + [(5-13h)(1-N_s) + (4+j)N_s]n}{13hN_sn + (4jh+5h)N_s} \\ &+ \frac{j(N_s-1)(4h-1) + (5h-1)(N_s+1) - 3N_sn^3}{13hN_sn + (4jh+5h)N_s} \times 100\% \end{aligned} \tag{4.54}$$

Where t_{PF} is the processing time per measurement for the particle filter and t_{ESP} is the same for the ESP-PF, h is the number of step size, n is the dimension of the state vector, and j denotes the computation complexity to evaluate the system dynamics.

4.6 NONLINEARITY AND CHOICE OF ESTIMATOR

In general, it is recommended to use the EKF for a mildly nonlinear system and the UKF performs better for highly nonlinear systems [8–10]. However, the quantitative relationship between the degree of nonlinearity and the performance of the UKF relative to the EKF has not been established.

Several methods for quantifying the nonlinearity of dynamic systems have been developed. In [32] the degree of nonlinearity of a nonlinear function is defined as the infimum norm of the deviation due to the best linear approximation of the function. The purpose of this measure of nonlinearity was to understand the linearizing effect of feedback control. Helbig et al. proposed a normalized version of this measure of nonlinearity [33]. Normalization of the measure of nonlinearity facilitated the comparison between different nonlinear systems. However, this work was focused on control-related analysis at different operating points. In [34] and [35, 36] a gap metric between a linear approximation of the nonlinear function and another suitably chosen linear function was defined as the measure of nonlinearity. This gap metric was utilized to study the robustness of feedback stability. Bate et al. suggested a geometric curvature-based nonlinearity measure [37]. In [38] this curvature-based measure of nonlinearity was utilized for the performance analysis of a maximum likelihood estimator for parameter estimation of a two-dimensional curve. However, this approach does not provide any generalized relation of nonlinearity with filter performance. Junkins [39, 40] suggested a measure of nonlinearity called the nonlinearity index, expressed as the supremum of the Hilbert-Schmidt (HS) norm of the change in the state transition matrix, normalized by the HS norm of the true state transition matrix. Necessarily, this nonlinearity index is the maximum possible normalized change in the second-order Taylor series term at a given time due to a deviation in initial conditions and was used to decide the suitable coordinate system for the satellite attitude estimation problem [40].

Recently, Liu et al. [41] proposed a measure of nonlinearity based on the norm of the deviation of the nonlinear function from the linear model. In this method, the minimum norm of the deviation considering a family of linear approximate functions is defined as the measure of nonlinearity. Using this measure of nonlinearity, it was demonstrated that, for a two-dimensional radar tracking problem, the performance of the UKF is better in the Cartesian coordinate system than in the polar coordinate system due to the higher degree of nonlinearity in polar coordinates.

4.6.1 Nonlinearity Index

As discussed previously, Junkins' definition of the nonlinearity index has a close relation with the second-order Taylor series terms. Hence, for system and measurement sets which are dominantly nonlinear up to at least the second-order Taylor series terms, there exists an explicit relation between the difference in the EKF and the UKF state estimation error and the nonlinearity indices of the system and the measurement models.

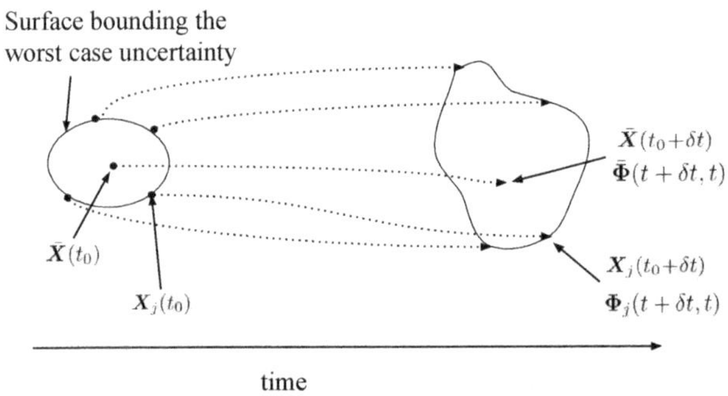

Figure 4.4: Different trajectories of the state vectors due to worst case initial conditions in two-dimensional space.

Consider a system and the corresponding measurement model described by (4.1) and (4.2). To compute the nonlinearity index, let us consider the expected state vector at time t_0, $\bar{\boldsymbol{X}}(t_0)$. The initial worst-case uncertainty of the state vector can be represented as a hyper-ellipsoid around $\bar{\boldsymbol{X}}(t_0)$. Select N uniformly distributed vectors $\boldsymbol{X}_j(t_0)$ $(j = 1, 2, ...N)$ on the hyper-ellipsoid. For these N initial state vectors, there will be N separate trajectories over time. This is explained graphically in Figure 4.4 for the two-dimensional case.

Considering sufficiently small time interval δt and $\frac{\partial \boldsymbol{f}}{\partial \boldsymbol{X}}$ remains unchanged within this interval, the state transition matrices $\boldsymbol{\Phi}_j(t+\delta t, t)$, $\bar{\boldsymbol{\Phi}}(t+\delta t, t)$ are defined as

$$\boldsymbol{\Phi}_j(t + \delta t, t) = e^{\frac{\partial \boldsymbol{f}}{\partial \boldsymbol{X}}\big|_{\boldsymbol{X}_j(t)}\delta t}$$
$$\bar{\boldsymbol{\Phi}}(t + \delta t, t) = e^{\frac{\partial \boldsymbol{f}}{\partial \boldsymbol{X}}\big|_{\bar{\boldsymbol{X}}(t)}\delta t}$$

The Jacobians of the measurement vector are defined as,

$$\frac{\partial \boldsymbol{h}(t+\delta t)}{\partial \boldsymbol{X}_j(t)} = \boldsymbol{H}_i(t+\delta t, t)$$
$$\frac{\partial \boldsymbol{h}(t+\delta t)}{\partial \bar{\boldsymbol{X}}(t)} = \overline{\boldsymbol{H}}(t+\delta t, t)$$

The nonlinearity index $\mathcal{N}_s$ of the system is defined as [39]

$$\mathcal{N}_s(t+\delta t, t) \triangleq \sup_{j=1,2\ldots,N} \frac{\|\boldsymbol{\Phi}_j(t+\delta t, t) - \bar{\boldsymbol{\Phi}}(t+\delta t, t)\|}{\|\bar{\boldsymbol{\Phi}}(t+\delta t, t)\|} \tag{4.55}$$

here $\|(.)\|$ is the Hilbert-Smith norm in an n-dimensional Hilbert space. Similarly, the nonlinearity index of the measurement vector $\mathcal{N}_m$ is

$$\mathcal{N}_m(t+\delta t, t) \triangleq \sup_{j=1,2\ldots,N} \frac{\|\boldsymbol{H}_j(t+\delta t, t) - \overline{\boldsymbol{H}}(t+\delta t, t)\|}{\|\overline{\boldsymbol{H}}(t+\delta t, t)\|} \tag{4.56}$$

4.6.2 Nonlinearity and Estimator Performance

Consider a nonlinear stochastic estimation problem given by (4.1) and (4.2) and the EKF and the UKF are separately used to estimate the states. Consider that the following assumptions are true:

1. The HS norm of the estimation error of the UKF $\epsilon_U(t)$ and that of EKF $\epsilon_E(t)$ are exponentially bounded
2. The system and measurement functions have dominant second-order nonlinearity; that is, $\|\boldsymbol{D}^2_{\Delta X}\boldsymbol{F}\| \gg \|\boldsymbol{D}^i_{\Delta X}\boldsymbol{F}\|$ and $\|\boldsymbol{D}^2_{\Delta X}\boldsymbol{h}\| \gg \|\boldsymbol{D}^i_{\Delta X}\boldsymbol{h}\|$ for $i = 3,\ 4,\ 5, \ldots\ n$

Then at a sufficiently long time t from the initialization of both estimators the following inequality will be true [42]:

$$\boxed{\begin{aligned}&\frac{\epsilon_E(t+\delta t) - \epsilon_U(t+\delta t)}{\epsilon_E(t+\delta t)}\\ &\leq min\left(\frac{1}{2}\left[\mathcal{N}_s(t+\delta t, t) + \mathcal{N}_m(t+\delta t, t)\right] + \zeta, 1\right)\end{aligned}} \tag{4.57}$$

where $\zeta < 1$. Inequality (4.57) predicts the maximum possible performance improvement by the UKF over the EKF based on nonlinearity indices of the system and observation models; however, it is possible that the performance improvement using the UKF is less than the maximum value for a given estimation problem. Since estimation errors are stochastic in nature, a lower and upper bound of the performance improvement by the UKF for a given probability would provide more decisive information. Now let us define the relative performance improvement of the UKF as

$$\gamma = \frac{\epsilon_E - \epsilon_U}{\epsilon_E} \tag{4.58}$$

and

$$\lambda = min\left(\frac{1}{2}\left[\mathcal{N}_s + \mathcal{N}_m\right] + \zeta, 1\right) \tag{4.59}$$

It can be shown that if the probability that γ remains within a certain range is $\mathbb{P}$ and $P < \mathbb{P} < 1$, then [42]

$$\boxed{\frac{\lambda}{P\sqrt{P}}\left[2(1+\sqrt{1-P})(\sqrt{P}-1)+P\right] < \gamma < \frac{\lambda}{\sqrt{P}}} \tag{4.60}$$

The nonlinearity index of various types of space vehicles and corresponding navigation performance improvement by the UKF relative to the EKF is shown in Figure 4.5 [42]. The performance of the UKF increases with increased nonlinearity and γ for each space vehicle type remains within the bound predicted by (4.60). Since the SPUKF and ESPUKF use approximation, the bound does not apply to these filters.

4.6.3 Choosing the Right Estimation Algorithm

The choice of a suitable state estimator for a given system is an important task for a designer of a navigation system. Section 4.6.2 provided a framework for choosing between the EKF and UKF quantitatively. The procedure to assess whether the UKF would provide significantly better estimation performance than the EKF for a given nonlinear system and measurement set is provided in Algorithm 4.1. One needs to decide the expected minimum probability P and a performance improvement threshold γ_{th} for the UKF to apply this algorithm. For example, we could check if the relative performance improvement of the UKF for a given system and measurement is above $\gamma_{th} = 0.4$ (i.e., 40% improvement compared to the EKF) with minimum probability 0.8 using Algorithm 4.1 for selecting the appropriate estimator.

It is recommended that a designer of a Kalman filter should apply this procedure to decide whether the EKF or the UKF is suitable for the system and

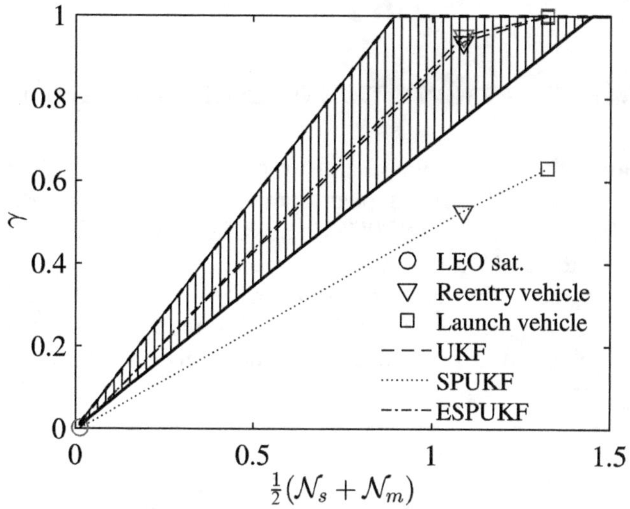

Figure 4.5: Combined nonlinearity index vs. relative UKF efficiency.

Algorithm 4.1 Algorithm for Selecting Kalman Filters

Compute $\frac{1}{2}(\mathcal{N}_s + \mathcal{N}_m)$ for the given system and measurement set
Compute λ: $\lambda = min\left(\frac{1}{2}\left(\mathcal{N}_s + \mathcal{N}_m\right), 1\right)$
Select the expected minimum probability P
Compute the lower bound (a) of γ: $a = \frac{\lambda}{P\sqrt{P}}\left[2(1+\sqrt{1-P})(\sqrt{P}-1)+P\right]$
if $a > \gamma_{th}$ **then**
    Use the UKF
else
    Use the EKF
end if

---

measurements. Note that there is no similar framework for selecting between particle filters and other estimators mentioned here. Further analysis is required to establish such a framework for the particle filter.

4.7 MATLAB® IMPLEMENTATION OF ESTIMATION ALGORITHMS

Having discussed the theoretical backgrounds of various estimation algorithms we will now discuss the MATLAB implementation of these algorithms. The MATLAB codes for the EKF, UKF, SPUKF, ESPUKF, particle filter, and ESP-PF are available in the GitHub repository `https://github.com/sbiswas8/space-nav-track`. The codes are written in a modular fashion so that the user can easily modify the codes to suit their needs. All the Kalman filter-based algorithms are implemented as independent functions, which take three structures as inputs. The general function calling syntax is as follows:

```
[state_c, rsd] = estimator_name(state, msmt, model)
```

Here, both `state` and `state_c` are the instances of the `state` structure. The `state` structure contains the following fields:

- `state.X`: The state vector
- `state.P`: The state covariance matrix

Essentially the `state` structure contains the previous state estimate and the state covariance matrix. The `state_c` structure contains the corrected state estimate and the state covariance matrix at the current measurement epoch. The `msmt` structure contains the following fields:

- `msmt.Z`: The measurement vector
- `msmt.dt`: The measurement time interval

The `model` structure contains four function handles:

- `model.dynamics`: Contains the state dynamics $\boldsymbol{f}(t, \boldsymbol{X}(t))$
- `model.Fjacobian`: The Jacobian of the state dynamics
- `model.msmtf`: Contains the measurement function $\boldsymbol{h}(\boldsymbol{X}(k))$
- `model.Mjacobian`: The measurement Jacobian

The output variable `rsd` is the residual vector. Note that the process and measurement noise covariance are not passed as inputs to the functions. These matrices should be defined as global variables in the MATLAB workspace. The user can also modify the codes to pass these matrices as inputs to the functions. We will now describe the EKF, UKF, SPUKF, and ESPUKF implementations in the subsequent parts of this section.

4.7.1 EKF Implementation

The EKF is implemented in MATLAB using the `ekf.m` function. The function is provided below:

```
function [state_c, rsd] = ekf(state, msmt, model)
dt = msmt.dt;
state_p = KalmanPrdctState(state.X,dt,model.dynamics);
P_p = KalmanPrdctCov(state.X,state.P,dt,model.Fjacobian);
[H, rsd] = PrdctMsmtCrtH(state_p,msmt,model.msmtf,model.
        Mjacobian);
state_C = KalmanCrct(state_p,P_p,rsd,H);
```

In the above function we have used four other functions:

- `KalmanPrdctState.m`: This function computes the predicted state vector using the state dynamics function handle provided in the `model` structure.
- `KalmanPrdctCov.m`: This function computes the predicted state covariance matrix using the Jacobian of the state dynamics function handle provided in the `model` structure.
- `PrdctMsmtCrtH.m`: This function computes the measurement Jacobian matrix and the residual vector using the measurement function handle provided in the `model` structure.
- `KalmanCrct.m`: This function computes the corrected state vector and the corrected state covariance matrix using Kalman's formulations described in the previous sections.

4.7.2 UKF Implementation

The UKF can be called using the `ukf.m` function. The function is provided as follows:

```
1  function [state_c, rsd] = ukf(state, msmt, model)
2  global Qk m n Wm0 Wm lambda_kf
3  m = length(state.X);
4  n = 2*m;
5  %% Parameters
6  alpha = sqrt(3/n);
7  beta = 2;
8  Kai = 0;
9  lambda_kf = alpha^2*(n + Kai) - n;
10 Wm0 = lambda_kf/(n + lambda_kf);
11 Wm = 1/(2*(n  + lambda_kf));
12 %% Augmentation
13 X = [state.X;zeros(m,1)];
14 Px = [state.P zeros(m);...
15  zeros(m) Qk];
16 %% filter
17 dt = msmt.dt;
18 Xai = CrtSigmaPt(X,Px);
19 [X_priori, Xai_priori] = UKFPrdctState(Xai,dt,model.
           dynamics);
20 P_p = UKFPrdctCov(X_priori,Xai_priori);
21 [P_XZ, Pvv, rsd] = UKFPrdctMsmt(X_priori,Xai_priori,msmt,
           model.msmtf);
22 state_C = UKFCrct(X_priori,P_p,rsd,P_XZ,Pvv);
```

Here lines 6 — 11 compute the UKF parameters. Lines 13 — 15 augment the state vector and the state covariance matrix. The rest of the functions used in this function are described below:

- `CrtSigmaPt.m`: This function computes the sigma points using the augmented state vector and the augmented state covariance matrix.
- `UKFPrdctState.m`: This function propagates the sigma points using the state dynamics function handle provided in the `model` structure.
- `UKFPrdctCov.m`: This function computes the predicted state covariance matrix using the predicted sigma points.

- `UKFPrdctMsmt.m`: This function computes the measurements corresponding to the predicted sigma points using the measurement function handle provided in the `model` structure.
- `UKFCrct.m`: This function computes the corrected state vector and the corrected state covariance matrix.

4.7.3 SPUKF Implementation

The `spukf.m` function is the MATLAB implementation of the SPUKF. The function is provided below:

```
function [state_C, rsd] = SPUKF(state,msmt,model)
global Qk m n Wm0 Wm lambda_kf
m = length(state.X);
n = 2*m;
%% Parameters
alpha = sqrt(3/n);
%beta = 2;
Kai = 0;
lambda_kf = alpha^2*(n + Kai) - n;
Wm0 = lambda_kf/(n + lambda_kf);
Wm = 1/(2*(n  + lambda_kf));
%% Augmentation
X = [state.X;zeros(m,1)];
Px = [state.P zeros(m);...
 zeros(m) Qk];
%% filter
dt = msmt.dt;
Xai = SPUKFCrtDelSig(X,Px);
[X_priori, Xai_priori] = SPUKFPrdctState(Xai,dt,model.dynamics,model.Fjacobian);
P_p = SPUKFPrdctCov(X_priori,Xai_priori);
[P_XZ, Pvv, rsd] = SPUKFPrdctMsmt(X_priori,Xai_priori,msmt,model.msmtf);
state_C = SPUKFCrct(X_priori,P_p,rsd,P_XZ,Pvv);
```

This implementation is very similar to the UKF implementation described earlier except for the `SPUKFPrdctState` function which implements the single propagation technique described in Section 4.4.4. Let us take a look at the `SPUKFPrdctState` function:

```
1  function [X_priori, Xai_priori] = SPUKFPrdctState(Xai,dt,dynamics,FJacobian)
2  global t_step m n Wm0 Wm
3  tspan = [0 dt];
4  %% Propagation of Sigma points
5  Xai_priori = nan(n/2,2*n+1);
6  [tout, Xpro] = runge_kutta4(dynamics,tspan,Xai.X,t_step);
7  Xai_priori(:,1) = Xpro(end,1:m)';
8  J = FJacobian(Xai.X(1:m,1));
9  PHI = expm(J*dt);
10 for count = 2:n+1
11  Xai_priori(:,count) = Xai_priori(:,1) + PHI*Xai.SigMat(1:m,count-1) + Xai.SigMat(m+1:n,count-1)*dt;
12 end
13
14 for count = n+2: 2*n+1
15  Xai_priori(:,count) = Xai_priori(:,1) - PHI*Xai.SigMat(1:m,count-1-n) - Xai.SigMat(m+1:n,count-1-n)*dt;
16 end
17 X_priori = Wm0*Xai_priori(:,1);
18 for count = 2:2*n+1
19  X_priori = X_priori + Wm*Xai_priori(:,count);
20 end
```

Here, the a posteriori state vector is propagated in line 6 using the Runge-Kutta fourth order method. The propagated state vector is then used to compute the Jacobian of the state dynamics function. The Jacobian is then used to compute the state transition matrix Φ, which is then used to compute the propagated sigma points in lines 10 — 16. The rest of the part of the function is similar to the UKF implementation.

4.7.4 ESPUKF Implementation

The `espukf.m` function is the MATLAB implementation of the ESPUKF. The function is provided below:

```
function [state_C, rsd] = ESPUKF(state,msmt,model)
global Qk m n Wm0 Wm lambda_kf
m = length(state.X);
n = 2*m;
%% Parameters
alpha = sqrt(3/n);
%beta = 2;
Kai = 0;
lambda_kf = alpha^2*(n + Kai) - n;
Wm0 = lambda_kf/(n + lambda_kf);
Wm = 1/(2*(n  + lambda_kf));
%% Augmentation
X = [state.X;zeros(m,1)];
Px = [state.P zeros(m);...
 zeros(m) Qk];
%% filter
dt = msmt.dt;
Xai = ESPUKFCrtDelSig(X,Px);
[X_priori, Xai_priori] = ESPUKFPrdctState(Xai,dt,model.
        dynamics,model.Fjacobian);
P_p = ESPUKFPrdctCov(X_priori,Xai_priori);
[P_XZ, Pvv, rsd] = ESPUKFPrdctMsmt(X_priori,Xai_priori,
        msmt,model.msmtf);
state_C = ESPUKFCrct(X_priori,P_p,rsd,P_XZ,Pvv);
```

As mentioned in Section 4.4.4 the ESPUKF uses the ESPT to propagate the sigma points. The `ESPUKFPrdctState` function implements the ESPT in the above function. The function is provided below:

```
function [X_priori, Xai_priori] = ESPUKFPrdctState(Xai,dt,
        dynamics,FJacobian)
global t_step m n Wm0 Wm
```

```
tspan = [0 dt];
%% Propagation of Sigma points
Xai_priori = nan(n/2,2*n+1);
[tout, Xpro] = runge_kutta4(dynamics,tspan,Xai.X,t_step);
Xai_priori(:,1) = Xpro(end,1:m)';
J = FJacobian(Xai.X(1:m,1));
PHI = expm(J*dt);
for count = 2:n+1
 XaiN1 = Xai_priori(:,1) + PHI*Xai.SigMat(1:m,count-1);
 delf = FJacobian(Xai.X(1:m,1) + Xai.SigMat(1:m,count -1)/2);
 delF = expm(delf);
 XaiN2 = Xai_priori(:,1) + PHI*Xai.SigMat(1:m,count-1)/2 +...
   delF*Xai.SigMat(1:m,count-1)/2;
 Xai_priori(:,count) = 2*XaiN2 - XaiN1 + Xai.SigMat(m+1:n, count - 1)*dt;
end

for count = n+2: 2*n+1
 XaiN1 = Xai_priori(:,1) - PHI*Xai.SigMat(1:m,count-1-n);
 delf = FJacobian(Xai.X(1:m,1) - Xai.SigMat(1:m,count -1-n)/2);
 delF = expm(delf);
 XaiN2 = Xai_priori(:,1) - PHI*Xai.SigMat(1:m,count-1-n)/2 -...
   delF*Xai.SigMat(1:m,count-1-n)/2;
 Xai_priori(:,count) = 2*XaiN2 - XaiN1 - Xai.SigMat(m+1:n, count - 1-n)*dt;
end
X_priori = Wm0*Xai_priori(:,1);
for count = 2:2*n+1
 X_priori = X_priori + Wm*Xai_priori(:,count);
end
```

Here, lines 10 — 26 implement the ESPT. The rest of the function is similar to the UKF implementation.

4.7.5 Particle Filter Implementation

The input-output structures in the particle filter and ESP-PF implementation are marginally different than the Kalman filter implementations described above. For both particle filters the function calling syntax is as follows:

```
[PF_out,msgl,N_eff] = estimator(PF_in,msmt,model,
        Sample_Info)
```

Here, the variables `PF_in` and `PF_out` are structures with the following fields:

- `PF_in`:
  - `PF_in.state`: The a posteriori state vector at the previous epoch
  - `PF_in.Pxx`: The a posteriori covariance matrix at the previous epoch
  - `PF_in.particles`: The particles at the previous epoch
  - `PF_in.Wt`: The weights of the particles at the previous epoch
- `PF_out`:
  - `PF_out.state`: The updated state vector at the current epoch
  - `PF_out.Pxx`: The updated covariance matrix at the current epoch
  - `PF_out.particles`: The particles at the current epoch
  - `PF_out.Wt`: The updated weights of the particles at the current epoch

The `model` structure has the following fields:

- `model.sys`: The function handle to the system dynamics function
- `model.meas`: The function handle to the measurement function
- `model.sys_noise`: The process noise covariance matrix
- `model.Pzx`: The likelihood function of the state $\boldsymbol{X}$ given measurement $\boldsymbol{Z}$

The `msmt` has the same structure described in Section 4.7. The remaining `Sample_Info` structure has the following fields:

- `Sample_Info.Ns`: The number of particles

- `Sample_Info.thld`: The threshold for resampling

The `particle_filter.m` is the MATLAB implementation of the particle filter. The function is provided below:

```
function [PF_out,msgl,N_eff] = particle_filter(PF_in,msmt, model,Sample_Info)
Ns = Sample_Info.Ns;
Nt = Sample_Info.thld*Ns;
dt = msmt.dt;
[particles,msgl] = CrtParticles(PF_in.particles,model.sys, Ns,dt);
[weights, N_eff] = CalcWeights(PF_in.Wt,PF_in.Pxx, particles,msmt,model.meas,model.Pzx,Ns);
prtcls_no_rsmpl = particles;
if N_eff < Nt   N_eff == inf
  [particles,weights] = resample_pf(particles,weights,Ns);
end

PF_out = PFupdate(particles,weights,msmt,model.meas,Ns);
PF_out.particles = particles;
PF_out.prtcls_n_rsmpl = prtcls_no_rsmpl;
```

The function `CrtParticles` creates the particles for the current epoch. The function `CalcWeights` calculates the weights of the particles. The function `resample_pf` resamples the particles if the effective number of particles is less than the threshold. The function `PFupdate` updates the state vector and the covariance matrix using the particles and the weights.

4.7.6 ESP-PF Implementation

The ESP-PF implementation uses the same structures as described above. However, the ESP-PF uses ESPT to propagate the particles. The function is provided below:

```
function [PF_out,msgl,N_eff] = ESP_PF(PF_in,msmt,model, Sample_Info)
Ns = Sample_Info.Ns;
Nt = Sample_Info.thld*Ns;
```

```
dt = msmt.dt;
[particles,msgl] = ESPCrtParticles(PF_in.state,PF_in.Pxx,
        model.sys,model.sysjacobian,Ns,dt);
[weights, N_eff] = CalcWeights(PF_in.Wt,PF_in.Pxx,
        particles,msmt,model.meas,model.Pzx,Ns);
prtcls_no_rsmpl = particles;
if N_eff < Nt  N_eff == inf
 [particles,weights] = resample_pf(particles,weights,Ns);
end
PF_out = PFupdate(particles,weights,msmt,model.meas,Ns);
PF_out.prtcls_n_rsmpl = prtcls_no_rsmpl;
PF_out.particles = particles;
```

Here, `ESPCrtParticles` is the function that creates the particles using ESPT. The remaining functions are the same as in the particle filter implementation.

4.8 SUMMARY

In this chapter we have posed the space vehicle navigation problem as an estimation problem, described the least square estimation, and then presented the Kalman filter, the EKF, and the UKF. In addition, we have briefly discussed the particle filter and its variants.

Since the inception of the Kalman filter, an indispensable mathematical tool for space vehicle navigation, it has proved to be a very reliable and flexible estimation technique. Different variants of Kalman filters have been in use in a multitude of practical and critical applications due to their ease of implementation. A plethora of research has been performed to refine and improve the performance of this estimation technique in nonlinear applications. The UKF is one of the most recognized variants of the Kalman filter. However, the computation time of the UKF is very high compared to the EKF and it has not been very obvious when the UKF performs better than the EKF. We discussed some pertinent problems related to the UKF, often encountered by the designers of Kalman filters in this chapter.

We presented two computationally efficient variants of the UKF — the SPUKF and ESPUKF. Additionally, a quantitative framework for choosing a suitable estimation algorithm was discussed.

The particle filter performance is superior to the EKF or UKF predominantly because of the way the PDF is approximated in the particle filter. However, the computation time and resource requirement for the particle filter is very high compared to the EKF or UKF due to the requirement of a large number of sample propagation. We briefly discussed various types of particle filters that address the degeneracy and processing time issues.

We have also discussed the MATLAB implementation of the estimators discussed in the chapter. In the subsequent chapters, we will discuss the application of these estimators to the space vehicle navigation problem.

References

[1] M. D. Shuster, D. S. Pitone, and G. J. Bierman, "Batch Estimation of Spacecraft Sensor Alignments I. Relative Alignment Estimation," p. 28,

[2] R. E. Kalman, "A new approach to linear filtering and prediction problems," *Journal of Basic Engineering*, vol. 82, no. 1, pp. 35–45, 1960.

[3] G. L. Smith, "Multivariable Linear Filter Theory Applied to Space Vehicle Guidance," *Journal of the Society for Industrial and Applied Mathematics Series a Control*, vol. 2, no. 1, pp. 19–32, Jan. 1964.

[4] B. Friedland and I. Bernstein, "Estimation of the state of a nonlinear process in the presence of nongaussian noise and disturbances," *Journal of the Franklin Institute*, vol. 281, no. 6, pp. 455–480, 1966.

[5] H. Cox, "On the estimation of state variables and parameters for noisy dynamic systems," *IEEE Transactions on Automatic Control*, vol. 9, no. 1, pp. 5–12, Jan. 1964.

[6] M. Athans, R. Wishner, and A. Bertolini, "Suboptimal state estimation for continuous-time nonlinear systems from discrete noisy measurements," *IEEE Transactions on Automatic Control*, vol. 13, no. 5, pp. 504–514, Oct. 1968.

[7] M. NøRgaard, N. K. Poulsen, and O. Ravn, "New developments in state estimation for nonlinear systems," *Automatica*, vol. 36, no. 11, pp. 1627–1638, 2000.

[8] S. J. Julier and J. K. Uhlmann, "New extension of the Kalman filter to nonlinear systems," *Signal Processing, Sensor Fusion, and Target Recognition VI*, International Society for Optics and Photonics, vol. 3068, 1997, pp. 182–194.

[9] S. Julier, J. Uhlmann, and H. Durrant-Whyte, "A new method for the nonlinear transformation of means and covariances in filters and estimators," *IEEE Transactions on Automatic Control*, vol. 45, no. 3, pp. 477–482, Mar. 2000.

[10] S. Julier and J. Uhlmann, "Unscented Filtering and Nonlinear Estimation," *Proceedings of the IEEE*, vol. 92, no. 3, pp. 401–422, Mar. 2004.

[11] Y. Bar-Shalom, X. R. Li, and T. Kirubarajan, *Estimation With Applications to Tracking and Navigation: Theory Algorithms and Software*. John Wiley & Sons, 2004.

[12] F. E. Daum, "Exact Finite-Dimensional Nonlinear Filters," *IEEE Transactions on Automatic Control*, vol. 31, no. 7, p. 7, 1986.

[13] R. Van Der Merwe, "Sigma-point Kalman filters for probabilistic inference in dynamic state-space models," Ph.D. dissertation, Oregon Health & Science University, 2004.

[14] S. Julier, "The scaled unscented transformation," *Proceedings of the 2002 American Control Conference (IEEE Cat. No.CH37301)*, Anchorage, AK, USA: IEEE, 2002, 4555–4559 vol.6.

[15] R. Zanetti and K. J. DeMars, "Joseph Formulation of Unscented and Quadrature Filters with Application to Consider States," *Journal of Guidance, Control, and Dynamics*, vol. 36, no. 6, pp. 1860–1864, Nov. 2013.

[16] S. Julier, "The spherical simplex unscented transformation," *Proceedings of the 2003 American Control Conference, 2003.*, vol. 3, Denver, CO, USA: IEEE, 2003, pp. 2430–2434.

[17] G. Chang, "Loosely Coupled INS/GPS Integration with Constant Lever Arm using Marginal Unscented Kalman Filter," *Journal of Navigation*, vol. 67, no. 03, pp. 419–436, May 2014.

[18] S. K. Biswas, L. Qiao, and A. G. Dempster, "A Novel a Priori State Computation Strategy for the Unscented Kalman Filter to Improve Computational Efficiency," *IEEE Transactions on Automatic Control*, vol. 62, no. 4, pp. 1852–1864, Apr. 2017.

[19] S. K. Biswas, "Computationally Efficient Non-Linear Kalman Filters for On-Board Space Vehicle Navigation," Ph.D. dissertation, UNSW Sydney, Sydney, Australia, 2017.

[20] M. S. Arulampalam, S. Maskell, N. Gordon, and T. Clapp, "A Tutorial on Particle Filters for Online Nonlinear/Non-Gaussian Bayesian Tracking," *Ieee Transactions on Signal Processing*, vol. 50, no. 2, p. 15, 2002.

[21] S. Thrun, "Particle Filters in Robotics," p. 9,

[22] T. Zhang, C. Xu, and M.-H. Yang, "Learning Multi-Task Correlation Particle Filters for Visual Tracking," *IEEE Transactions on Pattern Analysis and Machine Intelligence*, pp. 1–1, 2018.

[23] B. Ristic, A. Farina, D. Benvenuti, and M. Arulampalam, "Performance bounds and comparison of nonlinear filters for tracking a ballistic object on re-entry," *IEE Proceedings - Radar, Sonar and Navigation*, vol. 150, no. 2, p. 65, 2003.

[24] Y. Cheng and J. L. Crassidis, "Particle filtering for attitude estimation using a minimal local-error representation," *Journal of guidance, control, and dynamics*, vol. 33, no. 4, pp. 1305–1310, 2010.

[25] A. Mashiku, J. Garrison, and J. R. Carpenter, "Statistical Orbit Determination using the Particle Filter for incorporating Non-Gaussian Uncertainties.," *AIAA/AAS Astrodynamics Specialist Conference*, Minneapolis, Minnesota: American Institute of Aeronautics and Astronautics, Aug. 2012.

[26] F. Daum and J. Huang, "Particle flow for nonlinear filters with log-homotopy," *SPIE Defense and Security Symposium*, O. E. Drummond, Ed., Orlando, FL, Apr. 2008, p. 696 918.

[27] Y. Li and M. Coates, "Particle Filtering With Invertible Particle Flow," *IEEE Transactions on Signal Processing*, vol. 65, no. 15, pp. 4102–4116, Aug. 2017.

[28] A. Sankaranarayanan, A. Srivastava, and R. Chellappa, "Algorithmic and Architectural Optimizations for Computationally Efficient Particle Filtering," *IEEE Transactions on Image Processing*, vol. 17, no. 5, pp. 737–748, May 2008.

[29] W. Yi, M. R. Morelande, L. Kong, and J. Yang, "A Computationally Efficient Particle Filter for Multitarget Tracking Using an Independence Approximation," *IEEE Transactions on Signal Processing*, vol. 61, no. 4, pp. 843–856, Feb. 2013.

[30] K. Berntorp and S. Di Cairano, "Tire-Stiffness and Vehicle-State Estimation Based on Noise-Adaptive Particle Filtering," *IEEE Transactions on Control Systems Technology*, pp. 1–15, 2018.

[31] S. K. Biswas and A. G. Dempster, "Approximating Sample State Vectors Using the ESPT for Computationally Efficient Particle Filtering," *IEEE Transactions on Signal Processing*, vol. 67, no. 7, pp. 1918–1928, Apr. 2019.

[32] C Desoer and Y.-T. Wang, "Foundations of feedback theory for nonlinear dynamical systems," *IEEE Transactions on Circuits and Systems*, vol. 27, no. 2, pp. 104–123, 1980.

[33] A Helbig and W Marquardt, "Nonlinearity measures: Definition, computation and applications," *Journal of Process Control*, p. 11, 2000.

[34] A. El-Sakkary, "The gap metric: Robustness of stabilization of feedback systems," *IEEE Transactions on Automatic Control*, vol. 30, no. 3, pp. 240–247, Mar. 1985.

[35] T. T. Georgiou and M. C. Smith, "Optimal robustness in the gap metric," *IEEE Transactions on Automatic Control*, vol. 35, no. 6, pp. 673–686, 1990.

[36] W. Tan, H. J. Marquez, T. Chen, and J. Liu, "Analysis and control of a nonlinear boiler-turbine unit," *Journal of Process Control*, vol. 15, no. 8, pp. 883–891, Dec. 2005.

[37] D. M. Bates and D. G. Watts, "Relative Curvature Measures of Nonlinearity," *Journal of the Royal Statistical Society. Series B (Methodological)*, vol. 42, no. 1, pp. 1–25, 1980.

[38] M Mallick, S Arulampalam, Yanjun Yan, and A Mallick, "Connection between differential geometry and estimation theory for polynomial nonlinearity in 2D," *2010 13th International Conference on Information Fusion*, Edinburgh: IEEE, Jul. 2010, pp. 1–8.

[39] J. L. Junkins, "Von Karman Lecture: Adventures on the Interface of Dynamics and Control," *Journal of Guidance, Control, and Dynamics*, vol. 20, no. 6, pp. 1058–1071, Nov. 1997.

[40] J. L. Junkins and P. Singla, "A Tutorial on Nonlinearity of Orbit and Attitude Dynamics," p. 71,

[41] Y. Liu and X. R. Li, "Measure of Nonlinearity for Estimation," *IEEE Transactions on Signal Processing*, vol. 63, no. 9, pp. 2377–2388, May 2015.

[42] S. K. Biswas, L. Qiao, and A. G. Dempster, "A quantified approach of predicting suitability of using the Unscented Kalman Filter in a non-linear application," *Automatica*, vol. 122, p. 109 241, Dec. 2020.

Chapter 5

Navigation and Tracking of Satellites

5.1 OVERVIEW

Both navigation and tracking of satellites deal with the estimation of position, velocity, and time (PVT). Both problems are closely related and often distinguished by where the measurements are done and where the decision is made. In the case of navigation, the measurements or observations are taken from the onboard sensors. These observations are then processed to estimate the position and velocity of the satellite. The solution is then used in the flight computer for other operational purposes. In the case of tracking, the measurements are taken from a remote location and then processed externally to estimate the position and velocity of the satellite. Most of the time tracking is performed from one or more ground stations. The tracking solutions are often postprocessed and used as a data product. Although the purpose, solution location, and accuracy levels are different for navigation and tracking, the underlying mathematical frameworks are the same.

In this chapter, we will discuss the implementation of the EKF for both satellite navigation and tracking problems. First, we will present the initial position and velocity computation methods required for initializing the sequential estimator. Then we will discuss the system model used in the EKF and the measurement Jacobian formulation for multi-GNSS observations and azimuth, elevation, and range observations.

We will examine two example problems — ground-based tracking of a GEO satellite and GNSS-based navigation of an LEO satellite and discuss factors affecting the estimation performance. We will also demonstrate MATLAB implementation of various estimation algorithms for GNSS-based navigation of a LEO satellite.

Precise orbit determination (POD), real-time onboard navigation, and other practical considerations will be discussed at the end of this chapter.

5.2 INITIAL POSITION AND VELOCITY COMPUTATION

Calculating the initial position and velocity of a satellite is often referred to as initial orbit determination (IOD). A good initial guess results in the fast convergence of a sequential estimator. There are various methods available for IOD. Laplace and Gauss' methods are popular for IOD using optical observations where only angles are known. Both methods with proofs have been discussed in detail in various literature [1]. The Laplace method iteratively computes the range of the satellite from multiple-angle observations. This method requires the approximation of the first and second-time derivatives of the direction vector from the ground to the satellite using Legendre interpolation. Gauss' method requires three-angle measurements and utilizes the property of planar orbit. This method works better when the two consecutive observation angle difference is less than 60° [1]. Escobal's double-r iteration [2] iteratively improves estimates for the first and last ranges, and it is also effective for a large angular difference in observations, unlike Gauss' method.

Gooding proposed another approach for orbit determination using three angles-only observations. The core idea is the same as previously mentioned methods (i.e., estimation of range). This method starts with an initial range guess and constructs two satellite position vectors using the observed angles and the ground station coordinates. These are used in Lambert's problem solver, which computes the velocity of the satellite using two position vectors. Using this velocity and position vectors the orbit is propagated to the epoch for the next observation and the difference between the predicted and actual observation is calculated. By adjusting the initial guess iteratively the difference is minimized to obtain the range solution [3]. Another notable method is Gibb's method, which determines the velocity of the satellite from three position vectors [1]. Herrick proposed an improvement on Gibb's method that uses Taylor series approximation to get the velocity when the observed position vectors are close to one another [1].

With the advancement in sensors, radar, and signal processing, the available observations are predominantly range in addition to angles. In some instances, for example, in GNSS-based navigation, the observations are only GNSS pseudorange or carrier range. In such cases, it is more logical to estimate the satellite's position under consideration for two or three consecutive epochs and then estimate the velocity. The position solution can be obtained using the LSE discussed in Chapter

3. The velocity can be resolved by solving Lambert's problem (see Appendix A, Section 4) or using Gibb's or Herrick-Gibb's method.

5.3 SATELLITE NAVIGATION AND TRACKING USING SEQUENTIAL ESTIMATION

Sequential estimation processes the observations on the fly and utilizes prior information regarding the satellite motion and how the state is related to the observation. This is not the case for batch or least squared methods. From the information-theoretic perspective, a sequential estimator provides better accuracy than a batch estimator due to the use of additional information. In this section, we will discuss the implementation of the celebrated EKF for satellite navigation and tracking problems. In the satellite navigation problem, we will estimate the position and velocity of an LEO satellite using multi-GNSS observations recorded onboard the satellite. In the satellite tracking problem, we will estimate the position and velocity of a GEO satellite using azimuth, elevation, and range observations from a fixed ground station.

In a general sense, the EKF requires the following information apart from the observation:

- Initial guess of the mean state vector
- Initial guess of the error covariance
- Process and measurement noise covariance
- System model (differential equation), the state transition matrix
- Measurement model and Jacobian of the measurement vector

A KF-base sequential estimation algorithm is provided in Algorithm 5.1.

Variants of the Kalman filter differ in the state and covariance propagation methods; for example, the EKF uses numerical integration of the satellite dynamics equation and covariance propagation is done using linearization. On the other hand, in the UKF, the state and covariance are propagated using the UT. The rest of the steps remain the same.

5.3.1 System Model

Newton's law of gravitation is the basis of the in-orbit dynamics of an artificial satellite. However, due to the nonuniform density and irregular shape of the Earth,

Algorithm 5.1 Sequential Estimation Using KF Framework

```
if t = 0 then
    Initialize the state vector and covariance matrix
else
    Propagate the state vector and covariance matrix
end if
if New observation available then
    Calculate the Kalman gain
    Update the state vector and covariance matrix using new observation
end if
```

the actual motion of the satellite deviates from the motion modeled considering a spherical Earth with a uniform density. The gravitational potential of the Earth is expanded using a Legendre polynomial with latitude and longitude as parameters [4]. This expansion contains the zonal terms that are the Legendre functions of the sine of the latitude of the satellite and the tesseral terms that are the functions of both the latitude and the longitude. The coefficients of the zonal terms are called zonal harmonics and are denoted by J_n. The acceleration is computed from this Legendre polynomial and converted into a Cartesian coordinate system. For precise positioning, a high-fidelity force model is utilized using Legendre polynomials of higher degree and order [5]. In these simulated example problems, only J_2, J_3, and J_4 zonal harmonics are considered in the force model of the EKF.

The state vector for the satellite motion is:

$$\boldsymbol{X}_{sat} = \begin{bmatrix} \boldsymbol{r} \\ \boldsymbol{v} \end{bmatrix} = \begin{bmatrix} x & y & z & v_x & v_y & v_z \end{bmatrix}^T \tag{5.1}$$

where $\boldsymbol{r}= \begin{bmatrix} x & y & z \end{bmatrix}^T$ and $\boldsymbol{v}= \begin{bmatrix} v_x & v_y & v_z \end{bmatrix}^T$ are the position and the velocity vector of the satellite in the ECI frame. Neglecting the effect of the Sun and the Moon, the acceleration vector $\ddot{\boldsymbol{r}}$ can be expressed as [1, 6]:

$$\ddot{\boldsymbol{r}} = \begin{bmatrix} -\frac{\mu_e x}{r^3}(1 + J_2 C_{1x} + J_3 C_{2x} + J_4 C_{3x}) \\ -\frac{\mu_e y}{r^3}(1 + J_2 C_{1y} + J_3 C_{2y} + J_4 C_{3y}) \\ -\frac{\mu_e z}{r^3}(1 + J_2 C_{1z} + J_3 C_{2z} + \frac{\mu_e}{r^2} J_3 C_{3z} + J_4 C_{4z}) \end{bmatrix} \tag{5.2}$$

where

$$
\begin{aligned}
C_{1x} =& \frac{3}{2}\left(\frac{R_e}{r}\right)^2\left(1-5\frac{z^2}{r^2}\right)\\
C_{2x} =& \frac{5}{2}\left(\frac{R_e}{r}\right)^3\left(3-7\frac{z^2}{r^2}\right)\frac{z}{r}\\
C_{3x} =& -\frac{5}{8}\left(\frac{R_e}{r}\right)^4\left(3-42\frac{z^2}{r^2}+63\frac{z^4}{r^4}\right)\\
C_{1y} =& \frac{3}{2}\left(\frac{R_e}{r}\right)^2\left(1-5\frac{z^2}{r^2}\right)\\
C_{2y} =& \frac{5}{2}\left(\frac{R_e}{r}\right)^3\left(3-7\frac{z^2}{r^2}\right)\frac{z}{r}\\
C_{3y} =& -\frac{5}{8}\left(\frac{R_e}{r}\right)^4\left(3-42\frac{z^2}{r^2}+63\frac{z^4}{r^4}\right)\\
C_{1z} =& \frac{3}{2}\left(\frac{R_e}{r}\right)^2\left(3-5\frac{z^2}{r^2}\right)\\
C_{2z} =& \frac{5}{2}\left(\frac{R_e}{r}\right)^3\left(6-7\frac{z^2}{r^2}\right)\frac{z}{r}\\
C_{3z} =& \frac{3}{2}\left(\frac{R_e}{r}\right)^2\\
C_{4z} =& -\frac{5}{8}\left(\frac{R_e}{r}\right)^4\left(15-70\frac{z^2}{r^2}+63\frac{z^4}{r^4}\right)
\end{aligned}
$$

$r = \sqrt{x^2+y^2+z^2}$, R_e, and μ_e are the mean radius of the Earth and the gravitational parameter of the Earth, respectively. The differential equation for the state vector is:

$$
\dot{\boldsymbol{X}}_{sat} = \begin{bmatrix}\dot{\boldsymbol{r}}\\ \dot{\boldsymbol{v}}\end{bmatrix} + \boldsymbol{\nu}_{sat}(t) = \begin{bmatrix}\boldsymbol{v}\\ \ddot{\boldsymbol{r}}\end{bmatrix} + \boldsymbol{\nu}_{sat}(t) \tag{5.3}
$$

where $\boldsymbol{\nu}_{sat}$ is the process noise vector.

5.3.2 Satellite Position and Velocity Estimation Using Azimuth, Elevation, and Range Observation

In this example, we will track a GEO satellite (SATCAT no. 40940) from a ground station located at $-30.31°$ latitude, $149.56°$ longitude in the WGS84 reference system. The altitude of the ground station is 236.87 m from the WGS84 reference ellipsoid. The altitude is assumed to be 0. The observation is azimuth, elevation and range from the ground station and recorded every 1 minute for 200 minutes. The relation between the satellite state vector and the ground-based observations was discussed in detail in Chapter 3. Here we will discuss the measurement Jacobian that is required in the EKF.

The partial derivative of range with respect to the position vector $\boldsymbol{r}$ of the satellite in the ECI frame can be written as:

$$\frac{\partial \rho}{\partial \boldsymbol{r}} = \frac{\boldsymbol{r}_{gs}}{r_{gs}} \boldsymbol{R}_{TF} \boldsymbol{R}_{FI} \tag{5.4}$$

Where, ρ is the range of the satellite from the ground station, $\boldsymbol{r}_{gs}$ is the position vector of the satellite from the ground station in the topocentric frame, $\boldsymbol{R}_{TF}$ is the rotation matrix from the ECEF frame to the topocentric frame, and $\boldsymbol{R}_{FI}$ is the rotation matrix from the ECI frame to the ECEF frame.

The partial derivative of the azimuth angle with respect to the position vector $\boldsymbol{r}$ of the satellite in the ECI frame can be written as:

$$\frac{\partial A}{\partial \boldsymbol{r}} = \begin{bmatrix} \frac{r_N}{r_E^2+r_N^2} & -\frac{r_E}{r_E^2+r_N^2} & 0 \end{bmatrix} \boldsymbol{R}_{TF} \boldsymbol{R}_{FI} \tag{5.5}$$

where, $[r_E \; r_N \; r_U]' = \boldsymbol{r}_{gs}$. The partial derivative of the elevation angle with respect to the position vector $\boldsymbol{r}$ of the satellite in the ECI frame can be written as:

$$\frac{\partial E}{\partial \boldsymbol{r}} = \begin{bmatrix} -\frac{r_E r_U}{r_{gs}^2\sqrt{r_E^2+r_N^2}} & -\frac{r_N r_U}{r_{gs}^2\sqrt{r_E^2+r_N^2}} & \frac{\sqrt{r_E^2+r_N^2}}{r_{gs}^2} \end{bmatrix} \boldsymbol{R}_{TF} \boldsymbol{R}_{FI} \tag{5.6}$$

Considering $\boldsymbol{Z} = [\rho \; A \; E]'$, the measurement Jacobian with respect to the state vector $\boldsymbol{X} = [\boldsymbol{r} \; \boldsymbol{v}]'$ is:

$$\frac{\partial \boldsymbol{Z}}{\partial \boldsymbol{X}} = \begin{bmatrix} \frac{\partial \rho}{\partial \boldsymbol{r}} & 0 & 0 & 0 \\ \frac{\partial A}{\partial \boldsymbol{r}} & 0 & 0 & 0 \\ \frac{\partial E}{\partial \boldsymbol{r}} & 0 & 0 & 0 \end{bmatrix} \tag{5.7}$$

The state vector is initialized from the TLE elements. The standard deviation in the

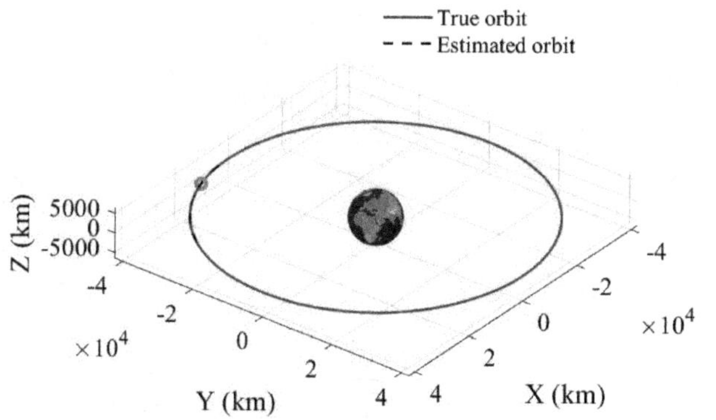

Figure 5.1: Estimated GEO orbit.

range is 2 km, and the same for azimuth and elevation are $1.06e-3°$. The estimated orbit in the ECI frame using the EKF for a single simulation is shown in Figure 5.1. Figure 5.2 shows the mean position estimation error using the designed EKF over 200 minutes for 20 Monte Carlo simulations. The sample standard deviation and the mean standard deviation computed by the EKF for these 20 runs are also provided. It can be observed that the sample standard deviation and the computed mean standard deviation are similar. The mean velocity errors in each axis, the mean standard deviation estimated by the EKF, and the sample standard deviation for 20 Monte Carlo simulations are shown in Figure 5.3. For both position and velocity estimation, the sample mean resides within the EKF estimated standard deviation and thus confirms the consistency of the EKF. Figure 5.4 shows the residual of measurements for a single run. The residuals are also unbiased, which further confirms the consistent operation of the estimator.

5.3.3 Satellite Position and Velocity Estimation Using Multi-GNSS Observation

In this example, we will estimate the position and velocity of an LEO satellite using simulated GPS and Galileo pseudorange and carrier range observations. The simulation procedure is out of the scope of this chapter, and details can be found

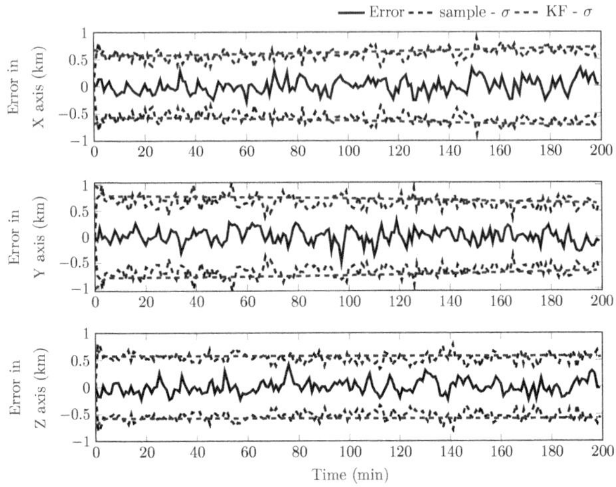

Figure 5.2: Position estimation error for the GEO satellite.

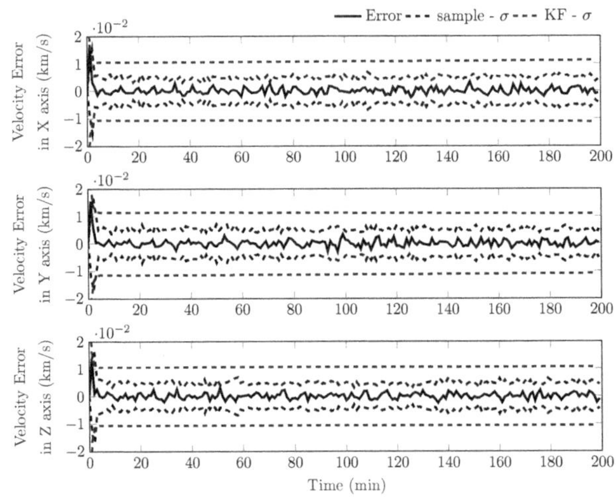

Figure 5.3: Velocity estimation error for the GEO satellite.

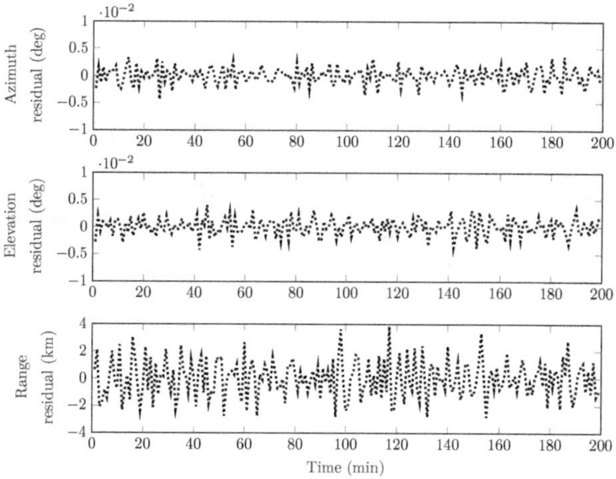

Figure 5.4: Residual.

in [7]. We will first discuss the augmentation required in the system model to accommodate multi-GNSS observations in the EKF framework and then discuss the observation processing and the measurement Jacobian formulation required in the EKF.

5.3.3.1 Augmented System Model

It should be noted that for GNSS-based navigation of satellites, the GNSS receiver clock bias is unknown. Hence, the receiver clock error must be estimated by including this bias in the state vectors [8]. For multi-GNSS applications, the intersystem clock bias of multiple navigation constellations must be considered in the measurement model and estimated. For example, if the GPS and Galileo constellations are used, then the receiver clock bias for Galileo is modeled as

$$\delta t_{GAL} = \delta t_{GPS} + \delta t_{ISB} \tag{5.8}$$

where δt_{GAL}, δt_{GPS}, and δt_{ISB} are the receiver clock bias for Galileo, the receiver clock bias for GPS, and the ISB, respectively.The receiver clock bias is modeled as a Markov process. The dynamics of the receiver clock bias vector can be represented

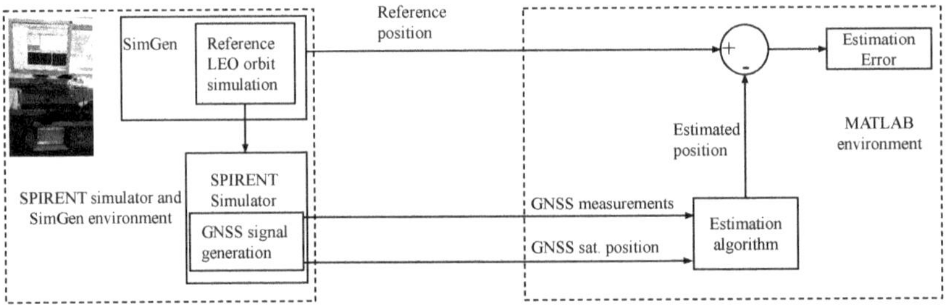

Figure 5.5: Simulation experiment with the LEO satellite.

as:

$$\begin{bmatrix} \dot{\delta t}_{GPS} \\ \dot{\delta t}_{ISB} \end{bmatrix} = \begin{bmatrix} 0 \\ 0 \end{bmatrix} + \begin{bmatrix} w_{GPS} \\ w_{ISB} \end{bmatrix} \tag{5.9}$$

where w_{GPS} and w_{ISB} are random noises. In this model, a zero-clock drift is assumed. For precision applications, drift is also included in the state vector. It should be noted that the clock bias and the ISB are very small quantities in terms of time, and using them in this form will result in an ill-posed problem. Hence δt_{GAL}, δt_{GPS}, and δt_{ISB} are multiplied by the speed of light and estimated as distance errors.

5.3.3.2 Multi-GNSS Measurement Simulation

A SPIRENT GNSS simulator with the SimGEN software is used to perform estimation experiments using GNSS observations with the LEO satellite. The SPIRENT GSS8000 GNSS simulator is capable of simulating GPS, GLONASS, and Galileo signals. This simulator takes vehicle motion as input from SimGEN application software and generates GNSS satellites' positions from the ephemeris and simulates the signals to be received by the user GNSS receiver from the visible satellites. The SimGEN is capable of simulating the tropospheric and ionospheric errors in measurements and also can incorporate receiver clock bias if the actual receiver is not available in the simulation.

The LEO satellite orbit is generated using the SimGEN software and the corresponding GPS and Galileo signals are generated by the SPIRENT GSS8000. The range observations from the simulator are used in the estimation algorithms for position and velocity estimation. The simulation process is shown in Figure 5.5.

5.3.3.3 Multi-GNSS Measurement Processing and Jacobian Formulation

Recall that navigation satellite position is required to estimate the position of the user satellite. Navigation satellites broadcast the ephemeris through the navigation message. The navigation satellite position can be calculated in the desired reference frame using the ephemeris. Since we have considered the position and velocity of the user satellite in the ECI frame, the navigation satellite position must be computed in the ECI frame in the EKF framework.

Using the GNSS measurement model described by (3.8) and considering the ionospheric error-free observation combination the partial derivatives of the GPS and Galileo observations with respect to the state vector are

$$\frac{\partial \rho_{i_{GPS}}}{\partial \boldsymbol{X}} = \begin{bmatrix} \frac{x_i - x_u}{r} & \frac{y_i - y_u}{r} & \frac{z_i - z_u}{r} & 0 & 0 & 0 & 1 & 0 \end{bmatrix} \tag{5.10}$$

$$\frac{\partial \rho_{i_{GAL}}}{\partial \boldsymbol{X}} = \begin{bmatrix} \frac{x_i - x_u}{r} & \frac{y_i - y_u}{r} & \frac{z_i - z_u}{r} & 0 & 0 & 0 & 1 & 1 \end{bmatrix} \tag{5.11}$$

The measurement Jacobian matrix can be formed by stacking these partial derivatives of the GPS and Galileo pseudorange measurements for each visible satellite. Figure 5.6 shows the position estimation error in each axis in the ECI frame and the associated standard deviations computed by the EKF. It can be observed that the errors are within the 1σ standard deviation. It should be noted that for both examples, the true state of satellites is known since these are numerically simulated examples. However, in practice, the true state of the satellites is not known and hence one needs to rely on the standard deviation to infer the quality of the estimation.

The same exercise can be performed using the UKF and the variants of the UKFs. The performance of the EKF, UKF, SPUKF, and ESPUKF are shown in Figure 5.7. It can be observed that all the estimators provide similar performance in this case. This behavior can be explained using the nonlinearity index discussed in Chapter 4. The performance of all the estimators is similar because, for the LEO satellite, the nonlinearity index is very small.

5.3.4 Factors Affecting Estimation Performance

In the context of GNSS-based satellite navigation, we have primarily focused on the estimation of position and velocity for satellites in circular orbits. However, it is also essential to evaluate the performance of different Kalman filters for satellites in elliptical orbits. To investigate this, we will examine the effectiveness of the EKF,

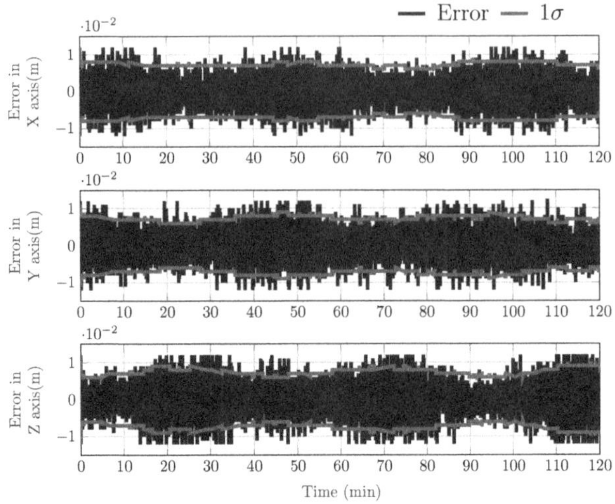

Figure 5.6: Orbit determination using multi-GNSS observation.

Table 5.1
Orbit Parameters

Orbit Parameter	Value
RAAN	-3.090^o
Argument of perigee	85.497^o
Inclination	97.445^o

UKF, SPUKF, and ESPUKF for the estimation of position and velocity in elliptical orbits.

For our experimentation, we have selected four orbits with different eccentricities: 0.1, 0.4, 0.7, and 0.9. In all cases, the perigees of the orbits are fixed at 400 km altitude from the mean sea level, ensuring consistency across the experiments. Additionally, all four orbits share the same orientation in space, meaning they lie in the same plane. The specific orbit parameters related to the orientation of these elliptical orbits can be found in Table 5.1.

During the numerical experiments, satellite navigation relies solely on the GPS constellation. The GPS constellation orbits and the selected orbits for our

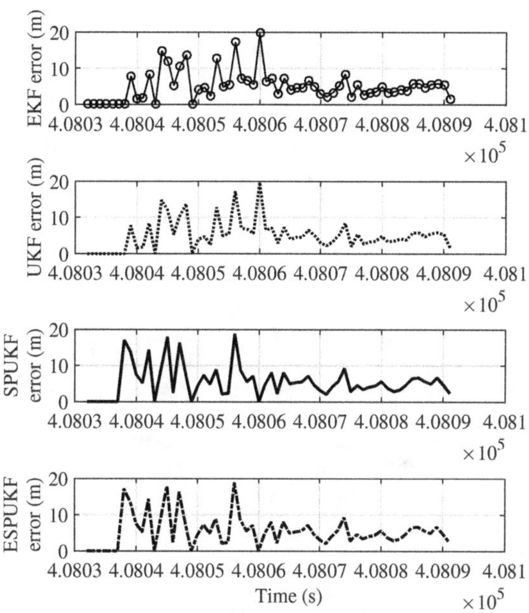

Figure 5.7: Comparison of the estimated position error.

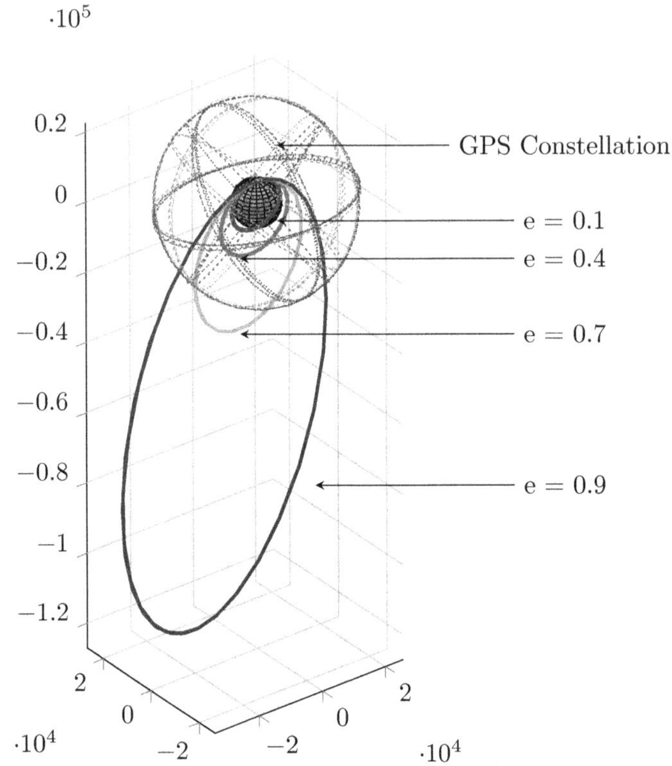

Figure 5.8: Test orbits.

experiments are visually depicted in Figure 5.8. It is clear from the figure that in the orbits with eccentricities of 0.1 and 0.4, the spacecraft remains within the space service volume. The space service volume refers to the region where reliable GPS signals are available for navigation purposes. Therefore, in these cases, the spacecraft maintains a consistent signal reception from the GPS satellites throughout its orbital trajectory.

However, for the orbits with eccentricities of 0.7 and 0.9, the spacecraft travels beyond the space service volume. This means that during certain portions of these orbits, the spacecraft may experience reduced or intermittent GPS signal availability, leading to potential challenges in navigation and estimation accuracy.

In the experimentation process, GPS pseudorange and range rate observations for the different orbits are generated using the method described in Section 5.3.3.2. The initial position of the spacecraft in each orbit is specified in the SIMGEN simulator, which then utilizes its internal orbit propagator to generate reference orbits and corresponding GPS observations. The generated GPS observation data consists of pseudorange and range rate measurements obtained from 12 channels at a rate of 1 Hz. These observations, along with the reference orbits, are recorded for all four orbits under investigation.

Subsequently, the recorded observation data is utilized independently in the EKF, SPUKF, and ESPUKF. Each filter is employed to estimate the position and velocity of the spacecraft for a duration of $1,000$ seconds.

A total of 100 estimation experiments are performed for each orbit using the EKF. An equal number of experiments are also conducted for the UKF, SPUKF, and ESPUKF. In each experiment, a different random noise sequence is added to the observation data. This helps to account for the stochastic nature of the measurements and assess the robustness of the filtering methods. The observation data, along with the added noise, is used as input to the filtering algorithms.

In order to accurately simulate the navigation satellite visibility in elliptical orbits, several factors are considered. The number of observations available at a given epoch depends on the number of visible satellites. However, the received signal power from a satellite also plays a crucial role. If the received signal power is too low, the receiver on the spacecraft may be unable to acquire or track the signal, resulting in the unavailability of pseudorange or range rate observations, even if the satellite is within line of sight of the spacecraft. Any observation associated with a signal power below -200 dB has been discarded to incorporate this practicality in the simulation. This ensures that only observations with a reasonable signal power are considered for estimation.

Figure 5.9 illustrates the number of visible navigation satellites with a received signal power above -200 dB for spacecraft in different orbits. It is worth noting that for orbits with eccentricities of 0.1, 0.4, and 0.7, there may be instances where the number of visible navigation satellites reduces below 8. However, for the orbit with an eccentricity of 0.9, the number of visible GPS satellites can reduce to 0, and in some epochs, 10 to 12 GPS satellites may be visible.

The received signal power for each visible GPS satellite can be computed by considering the transmitted signal power recorded from SIMGEN and accounting for the path loss. Figure 5.10 displays the variation in received signal power for the visible GPS satellites across all epochs for different orbits. From the figure, it is evident that orbits with higher eccentricities tend to exhibit very low received

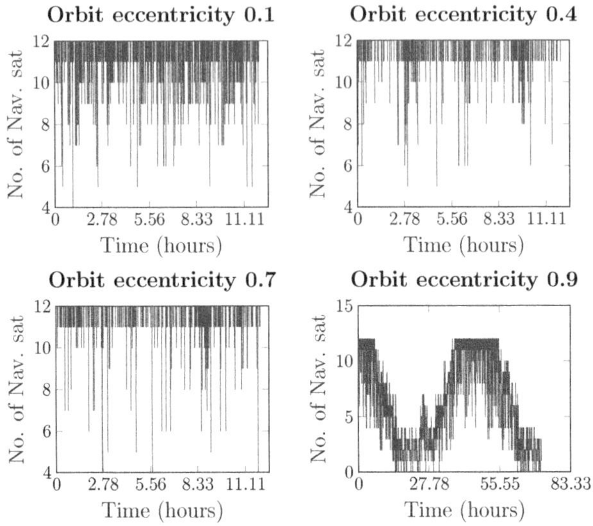

Figure 5.9: Satellite visibility for different orbit eccentricities.

signal power. This can pose challenges in practical applications where reliable signal reception is crucial. To mitigate this issue, it is recommended to employ high-gain GPS antennas, which can help improve signal reception and ensure more reliable navigation in orbits with higher eccentricities.

Let us now discuss the estimation performance of various estimation algorithms for each orbit. Figure 5.11 shows the root mean square (RMS) errors for all experiments conducted. For the 0.1 eccentricity orbit, it is surprising to observe that the EKF outperforms the UKF, which is an unusual outcome. However, both the SPUKF and the ESPUKF consistently exhibit lower estimation errors compared to the EKF and the UKF in all experiments.

Moving on to the 0.4 eccentricity orbit, the estimation error of the UKF becomes comparable to that of the EKF. However, the errors for the SPUKF and the ESPUKF remain lower than those of the EKF and the UKF.

In the case of the 0.7 eccentricity orbit, all the UKFs deliver better estimation accuracy than the EKF. This indicates that the UKFs are more suitable for this particular orbit.

For the 0.9 eccentricity orbit, the performance of the UKFs remains similar, with errors lower than those of the EKF. However, it is important to note an

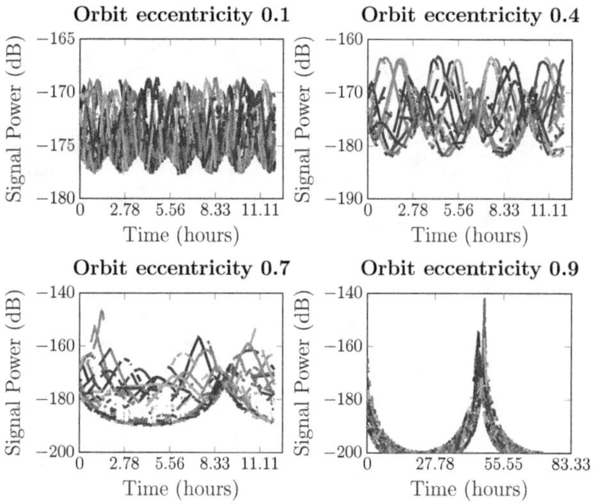

Figure 5.10: Signal power for different orbit eccentricities.

increased variation in estimation errors throughout the experiments. This can be attributed to the fact that in the 0.9 eccentricity orbit, the available GPS satellites are fewer compared to the 0.7 eccentricity orbit. Additionally, the position dilution of precision (PDOP) tends to be larger in most epochs, which degrades the performance of the estimation algorithms [9].

To further analyze the results, the variation in mean and standard deviation of RMS errors for experiments with different eccentricities is depicted in Figures 5.12 and 5.13. It is noteworthy that the estimation accuracies of all the UKFs decrease for the 0.9 eccentricity compared to the 0.7 eccentricity. This can be attributed to the reduced number of available GPS satellites in the 0.9 eccentricity orbit, as well as the larger PDOP values, which negatively impact the performance of the estimation algorithms [9].

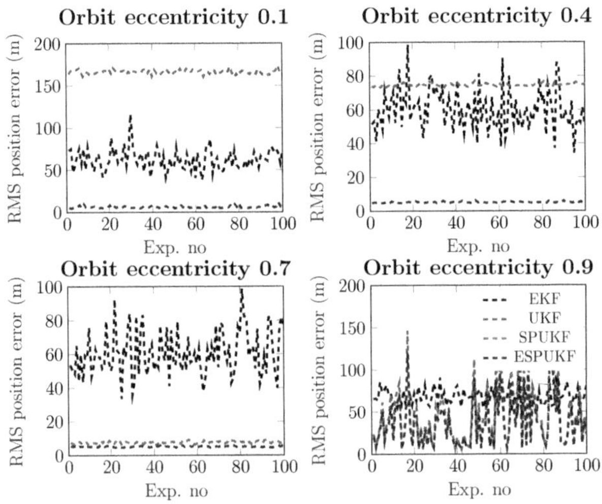

Figure 5.11: RMS errors for various algorithms.

Table 5.2

Performance of Carious Estimation Algorithms

Ecc.	0.1		0.4		0.7		0.9	
	Processing Time (ms)	Mean Position Error (m)	Processing Time (ms)	Mean Position Error (m)	Processing Time (ms)	Mean Position Error (m)	Processing Time (ms)	Mean Position Error (m)
EKF	10	61.870	10	59.901	10	60.643	10	67.980
UKF	278	166.106	276	74.729	281	7.565	275	45.099
SPUKF	40	5.978	40	5.088	40	5.104	39	37.295
ESPUKF	40	5.978	40	5.088	40	5.104	39	37.295

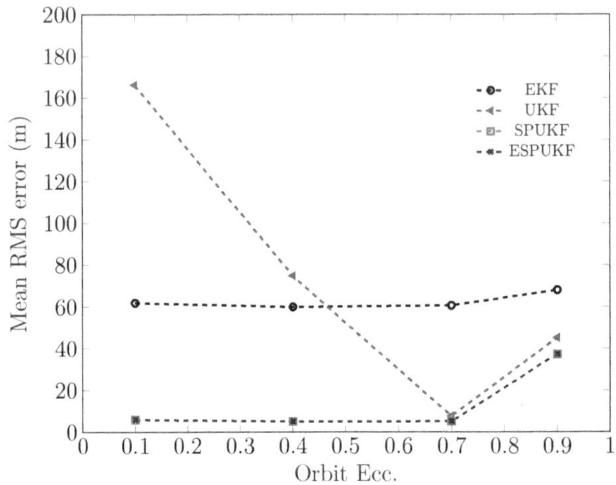

Figure 5.12: Mean of RMS errors for 100 runs.

Figure 5.14 shows the mean processing time per estimation step for the 100 experiments conducted for each orbit and estimation algorithm. It is observed that there is very minor variation in processing time for different eccentricities. The UKF requires the highest processing time, while the EKF has the lowest processing time. On the other hand, both the SPUKF and the ESPUKF exhibit significantly lower processing times compared to the UKF.

These results indicate that the SPUKF and ESPUKF offer a good balance between estimation accuracy and processing time, as they achieve lower estimation errors than the EKF and UKF while also requiring less computational effort. The UKF, while delivering comparable accuracy in some cases, comes at the expense of higher processing time. Therefore, the SPUKF and ESPUKF are favorable choices for efficient GNSS-based spacecraft navigation in elliptical orbits.

The results of the numerical experiments are summarized in Table 5.2. The SPUKF and the ESPUKF are better choices than the EKF and the UKF, for GNSS-based spacecraft navigation in elliptical and high earth orbits, considering the processing time and the consistency in estimation performance.

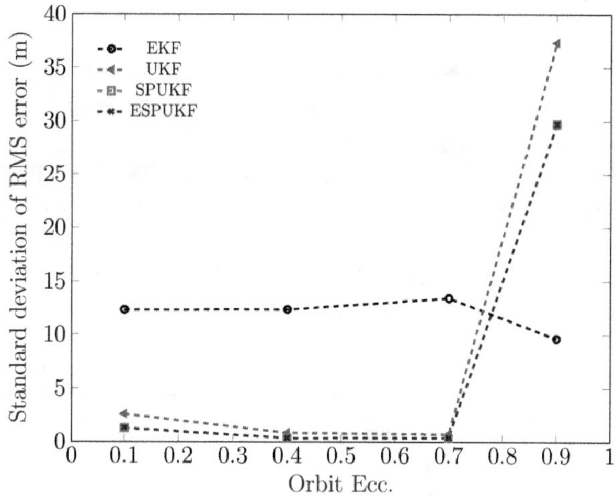

Figure 5.13: Standard deviation of RMS error for 100 runs.

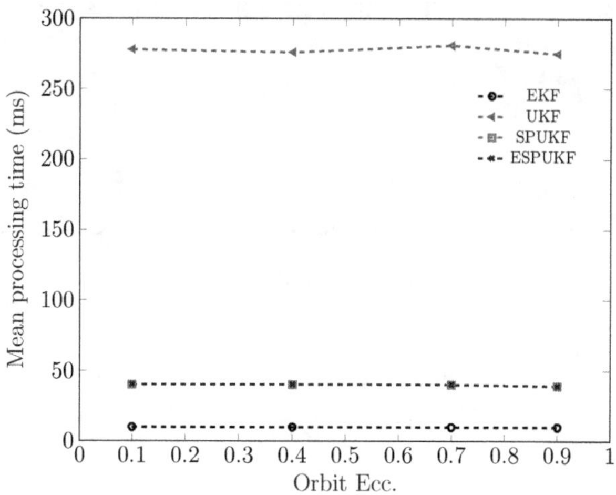

Figure 5.14: Mean execution time for different orbital eccentricities.

5.4 EXAMPLE MATLAB IMPLEMENTATION OF GPS-BASED SATELLITE NAVIGATION

We will now present an example MATLAB implementation of the GPS-based satellite navigation algorithm. The observation data used here are generated using the simulation method described in Section 5.3.3.2. The results discussed in the previous section were generated using this implementation. Here we will utilize the estimator functions described in Section 4.7. The code file and associated data files are available at `https://github.com/sbiswas8/space-nav-track`. This code can run the EKF, UKF, SPUKF, and ESPUKF based on the user's choice. The syntax for running the code is as follows:

```
[EST,m] = KF_sim(obs_data_adrs,SAT_INI,FilterType)
```

Here `obs_data_adrs` is the address of the data file, and `SAT_INI` is a structure that contains the initial deviation in the satellite state vector from the reference state vector, initial error covariance, process, and measurement noise information. `FilterType` is the type of filter to be used. The variable `EST` is a structure containing the estimated states and the covariance matrices. The variable `m` is a status message. Here we have presented the code in several code snippets for ease of explanation.

The first code snippet loads the data from the address stored in `obs_data-_adrs` variable. The stored data file is essentially a `.mat` file that contains the following variables:

- `GPSM` - GPS measurements
- `PVT_GPS` - Position and velocity of the GPS satellites
- `Ref_R_ECEF` - True position of the user satellite in ECEF frame
- `Ref_V_ECEF` - True velocity of the user satellite in ECEF frame
- `EPH` - Ephemeris data of the GPS satellites

The next lines select the estimator function based on the variable `FilterType`.

```
function [EST,m] = KF_sim(obs_data_adrs,SAT_INI,
    FilterType)
load(obs_data_adrs) % contains GPSM,PVT_GPS,Ref_R_ECEF,
    Ref_V_ECEF,EPH
switch FilterType
```

```
        case 'EKF'
            filter_alg = @EKF;
        case 'UKF'
            filter_alg = @UKF;
        case 'SPUKF'
            filter_alg = @SPUKF;
        case 'ESPUKF'
            filter_alg = @ESPUKF;
end
```

The subsequent lines of code initialize the necessary local and global variables.

```
%% constants and parameters
global c omega_earth R Qk sigma_a sigma_v GPSPVT theta
     t_step
c = 299792458; %m/s
omega_earth = [0; 0; 7.2921158553*10^-5];
sigma_R = SAT_INI.sigmar;
sigma_RR = SAT_INI.sigmarr;
L1 = 1575.42*10^6;
if strcmp(FilterType,'EKF') == 1
    Qk = SAT_INI.EKFQk;
else
    Qk = SAT_INI.UKFQk;
end
sigma_a = sqrt(1e-8);
sigma_v = sqrt(1e-8);
t_step = 0.01;
lambda = c/L1;
TIME = 1:SAT_INI.Time;
Ref_R_ECI = nan(length(TIME),3);
Ref_V_ECI = nan(length(TIME),3);
%% Computing initial position using LSQ
start_i = check_visible_sat(GPSM);
[X_lsq,del_rho, del] = lsq_satnav(GPSM(start_i),
     PVT_GPS(start_i), EPH);
```

```
    %% Converting reference data in ECI frame
    for i = start_i: length(TIME)
        THETA = sha_calc(GPSM(i).GPST);
        Ref_R_ECI(i,:) = ECEF2ECI(THETA, Ref_R_ECEF(i,:)')
          ';
        Ref_V_ECI(i,:) = ECEF2ECI(THETA,Ref_V_ECEF(i,:)')'
          + cross(omega_earth,Ref_R_ECI(i,:)')';
    end
```

The next snippet of code initializes the `model` structure described in Section 4.7. Note that the system model function passed in the `model` is different for the EKF and the UKF. This is because the UKF requires augmentation of the state vector.

```
    %% model handling
    if strcmp(FilterType,'EKF') == 1
        Sat_model.dynamics = @sat_dynamics;
    else
        Sat_model.dynamics = @sat_dynamicsUKF;
    end
    Sat_model.Fjacobian = @sat_jacobian;
    Sat_model.msmtf = @GPS_msmt;
    Sat_model.Mjacobian = @GPS_Mjacobian;
```

The following code snippet initializes the filter. The initial state vector and error covariance are assigned from the `SAT_INI` structure which is essentially generated using the LSE. The `RSDR` and `RSDV` variables are used to store the residuals of the position and velocity measurements, respectively.

```
    %% Filter initialization
    X0 = [Ref_R_ECI(1,:) Ref_V_ECI(1,:) X_lsq(4)]' +
         SAT_INI.del;
    STATE = struct([]);
    STATE(1).X = X0;
    STATE(1).P = diag(SAT_INI.STD.^2);
    STATE(1).STD = SAT_INI.STD;
    RSDR = [];
    RSDV = [];
```

The final code snippet shows how the various estimator functions are used to estimate the state vector in a loop and the necessary data are stored for analysis of the results.

```
%% EKF
profile on
msg1 = sprintf('%s processing started\n',FilterType);
fprintf(msg1);
m = numel(msg1);
for i = 1:length(TIME)-1
    GPSPVT = PVT_GPS(i+1);
    MSMT.PRN = GPSM(i+1).PRN;
    MSMT.Z = [GPSM(i+1).R;GPSM(i+1).RR];
    MSMT.dt = GPSM(i+1).dt;
    theta = sha_calc(GPSM(i+1).GPST);
    NS = length(GPSPVT.SP);
    GPSPVT.EPH = EPH;
    GPSPVT.PRN = GPSM(i+1).PRN;
    GPSPVT.GPST = GPSM(i+1).GPST;
    GPSPVT.R = GPSM(i+1).R;
    %% Constructing R
    R = zeros(2*NS);
    for k = 1:2*NS
        if k<= NS
            R(k,k) = sigma_R^2;
        else
            R(k,k) = sigma_RR^2;
        end
    end
    [state_C, rsd] = filter_alg(STATE(i),MSMT,
     Sat_model);
    STATE(i+1).X = state_C.X;
    STATE(i+1).P = state_C.P;
    STATE(i+1).STD = state_C.STD;
    RSDR = [RSDR; rsd(1:NS)];
    RSDV = [RSDV; rsd(NS+1:2*NS)];
```

```
        prcsd = i/length(TIME)*100;
        msg = sprintf('%s processing: %0.00f percent
          complete\n',FilterType,prcsd);
        %if rem(prcsd,10) == 0
        fprintf(repmat('\b',1,m));
        fprintf(msg);
        m = numel(msg);
        %end
    end
    PR = profile('info');
    for i = 1:length(STATE)
        EST.X_est(i,:) = STATE(i).X';
        EST.STD(i,:) = STATE(i).STD';
    end
    EST.extime = extcalc(PR,FilterType);
    lngth = length(EST.X_est);
    EST.error = [Ref_R_ECI(1:lngth,:) Ref_V_ECI(1:lngth,:)]
            - EST.X_est(:,1:6);
    EST.RSDR = RSDR;
    EST.RSDV = RSDV;
```

5.5 PRECISE ORBIT DETERMINATION

Precise orbit determination (POD) has a more stringent accuracy requirement than the examples shown earlier. POD demands accuracy often at the cm level. Generally, the POD solution is generated through the postprocessing of measurements. The first step in the POD process is to collect data from various sources, including ground-based tracking stations, onboard sensors, and other satellite systems. Typical observations are range, angle, or Doppler measurements. The sources that are used to obtain these observations are GNSS, optical telescopes, radar, VLBI, and SLR.

The raw data collected from various sources may contain errors, biases, and outliers that can affect the accuracy of the orbit determination. Therefore, the next step is to preprocess the data to remove the outliers and correct these errors and biases. This involves applying various correction models to the raw data, for example atmospheric delay correction, transmitter/receiver clock bias correction, navigation satellite position correction (for GNSS observation), relativistic corrections, and so forth. The POD using GNSS observation requires precise orbit information of the GNSS satellites, which comes from the International GNSS Services (IGS). In addition, these precise positions are available every 15 minutes, and hence the orbit must be interpolated accurately in between.

Once the preprocessed data is available, the orbit determination process starts. The state vector of interest includes the position and velocity of the satellite and various parameters like atmospheric drag coefficient, solar radiation pressure coefficient, receiver clock bias, and drift rate in case of GNSS-based observations. The EKF is used in this case as well; however, the system model used to propagate the state is more detailed, which includes gravitational harmonics of higher order and degree, solar radiation pressure, atmospheric drag, and tidal effects. After the sequential estimation, further smoothening is performed if required. The estimated orbit is then validated by comparing it with the actual observations made by ground-based tracking stations or other sensors. Any discrepancies between the estimated and actual orbits are investigated and corrected.

5.6 REAL-TIME ONBOARD NAVIGATION AND OTHER PRACTICAL CONSIDERATIONS

Modern LEO satellites are equipped with GNSS receivers for onboard navigation. It should be noted that onboard processing is constrained by computational resources and power limits. The navigation solution process must run continuously in the background of the flight computer and support other operations by providing PVT solutions within a stipulated time. Often design trade-offs are made in the navigation solution, especially in the state propagation method. The choice of numerical integrator for the EKF prediction step is governed by the required error tolerance and the required solution rate. For example, a fourth-order Runge-Kutta method may provide better numerical stability and accuracy than the Euler integration; however, the latter can be used if a high-rate solution is required.

As discussed in Chapter 4, the nonlinearity of a system influences the accuracy of Kalman filters. The satellite in an orbit with higher eccentricity naturally will have a higher change in the state transition matrix for small perturbation than that in an orbit with lower eccentricity. As a result, although the dynamics of a spacecraft in any orbit are the same, the nonlinearity will be different depending on the eccentricity of the orbit. This difference in the nonlinearity implies that the performance of various Kalman filters will be different in orbits with different eccentricities.

In the last decade, several experiments have been performed to assess the performance of GNSS-based navigation of satellites above the space service volume. Navigator GPS receiver developed by the NASA Goddard Space Flight Center and the SGR-GEO receiver developed by the Surrey Satellite Technology Ltd. (SSTL) demonstrated the effectiveness of GNSS-based navigation beyond the space service volume. However, navigation using GNSS observation above the space service volume comes with certain challenges. If the orbit's apogee is above the geostationary orbit (i.e., the spacecraft is in a high earth orbit (HEO)), then the state vector of the spacecraft motion may be only partially observable due to the limited availability and low signal power of GNSS signals. It has been observed that the performance of the EKF degrades with an increase in the eccentricity of the orbit [10]. The UKF provides better accuracy in the case of high eccentricity orbits. For onboard processing in these cases, the computationally efficient variants of the UKF, SPUKF, or ESPUKF can be used to reduce the computational resource usage.

References

[1] D. A. Vallado, *Fundamentals of Astrodynamics and Applications*. Springer Science & Business Media, 2001, vol. 12.

[2] P. R. Escobal, *Methods of Orbit Determination*. 1965.

[3] R. H. Gooding, "A new procedure for the solution of the classical problem of minimal orbit determination from three lines of sight," *Celestial Mechanics and Dynamical Astronomy*, vol. 66, no. 4, pp. 387–423, 1996.

[4] W. M. Kaula, "Statistical and harmonic analysis of gravity," *Journal of Geophysical Research*, vol. 64, no. 12, pp. 2401–2421, Dec. 1959.

[5] O. Montenbruck, P. Swatschina, M. Markgraf, S. Santandrea, J. Naudet, and E. Tilmans, "Precision spacecraft navigation using a low-cost GPS receiver," *GPS Solutions*, vol. 16, no. 4, pp. 519–529, Oct. 2012.

[6] L. Qiao, "A Multiple GNSS-based Orbit Determination Algorithm for Geostationary Satellites," p. 13,

[7] S. Biswas, L. Qiao, and A. Dempster, "Space-borne GNSS Based Orbit Determination using a SPIRENT GNSS Simulator," *Australian Space Research Conference*, Adileide, Australia, 2014, p. 7.

[8] E. D. Kaplan and C. Hegarty, Eds., *Understanding GPS: Principles and Applications* (Artech House Mobile Communications Series), 2nd ed. Boston: Artech House, 2006.

[9] S. K. Biswas, L. Qiao, and A. G. Dempster, "Effect of PDOP on performance of Kalman Filters for GNSS-based space vehicle position estimation," *GPS Solutions*, vol. 21, no. 3, pp. 1379–1387, Jul. 2017.

[10] S. K. Biswas and A. G. Dempster, "GNSS-based spacecraft navigation in elliptical and High Earth Orbits using Single Propagation Unscented Kalman Filters," *2019 Fifth Indian Control Conference (ICC)*, Jan. 2019, pp. 283–288.

Chapter 6

Navigation and Tracking During Launch and Reentry

6.1 OVERVIEW

Navigation and tracking are critical for launch and reentry missions. Accurate position and velocity information is required to control the launch vehicle during ascent and guide the vehicle to the desired orbit insertion point at the desired velocity. In addition, this information is required in the flight termination system (FTS) for range safety.

The reentry vehicle is tracked from the ground to initiate the reentry procedure at the desired reentry point, predict the landing site, and initiate various decelerating sequences for a soft landing. For a fully autonomous reentry vehicle, navigation is performed onboard, which is then used in the guidance and the control algorithm to bring the vehicle to the desired landing site.

For the preliminary design of navigation and tracking algorithms, one should create a numerical simulation scenario, implement the tracking and navigation algorithms in the simulation, and then analyze the estimation performance for tuning purposes. This chapter will discuss the estimator design process for

- Ground-based tracking of a reentry vehicle.
- GPS-based navigation algorithm design of a launch vehicle.

We will demonstrate performance of various estimators in the above-mentioned scenarios using numerical simulation MATLAB-based estimation algorithm implementation for both problems will also be discussed in this chapter.

6.2 GROUND-BASED TRACKING OF REENTRY VEHICLE

Ground-based reentry vehicle tracking is generally performed using radar observations. Radar provides azimuth, elevation, range, and range rate. In this numerical simulation example, it is assumed that the reentry trajectory is restricted in a 2D plane, and hence the azimuth angle is fixed. The reference trajectory is generated using the reentry vehicle dynamics described in Chapter 2. The process noise standard deviations of $10^{-4}m/s$ and 10^{-3} are considered in acceleration and the time derivative of the aerodynamic coefficient, respectively.

The range and elevation are considered as the observations in this example. The effect of the Earth's curvature is considered in the measurement simulation. The geometry of radar tracking for a reentry vehicle is shown in Figure 6.1.

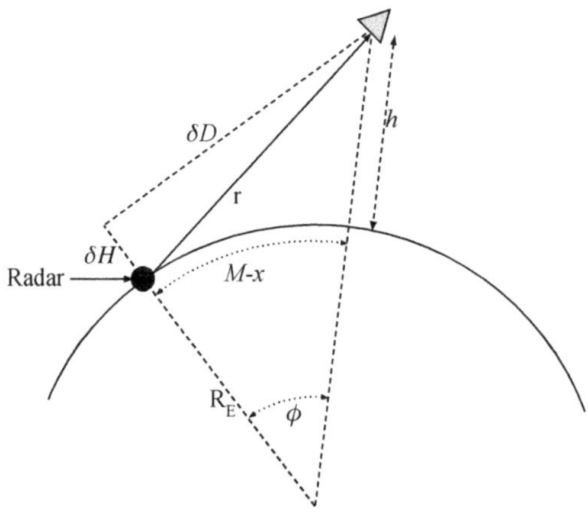

Figure 6.1: Radar tracking geometry for a reentry vehicle.

In the figure, r is the geometric distance of the reentry vehicle, M is the initial down-range distance of the radar from the reentry point, h is the altitude, and x is the downrange distance of the reentry vehicle from the reentry point. ϕ is the angular distance between the radar and the reentry vehicle and can be written as

$$\phi = \frac{M - x}{R_E} \tag{6.1}$$

Let

$$\delta H(t) = (R_E + h(t)) \cos \phi - R_E \tag{6.2}$$

$$\delta D(t) = (R_E + h(t)) \sin \phi \tag{6.3}$$

Then the range r and elevation E is modeled as

$$r(t) = \sqrt{\delta H(t)^2 + \delta D(t)^2} + \omega_r(t) \tag{6.4}$$

$$E(t) = \tan^{-1} \frac{\delta D(t)}{\delta H(t)} + \omega_E(t) \tag{6.5}$$

here, $\omega_r(t)$ and $\omega_E(t)$ are zero mean white Gaussian noise with a standard deviation of 20 m and 17.5 millirad, respectively. It should be noted that the additive white Gaussian noise assumption is reasonable for many applications [1]. However, the noise behavior may be different due to clutter or jamming, which is not considered in this analysis. In addition to the noise, biases in observations are expected in practice, due to radar calibration and measurement processes [1]. These can be estimated by considering these biases in the state vector. In this section, we will not consider bias in measurements and demonstrate the bias estimation technique in the next section.

The range and elevation measurements are generated using (6.4) and (6.5). The block diagram for the reentry vehicle trajectory estimation experiments using radar observations is shown in Figure 6.2.

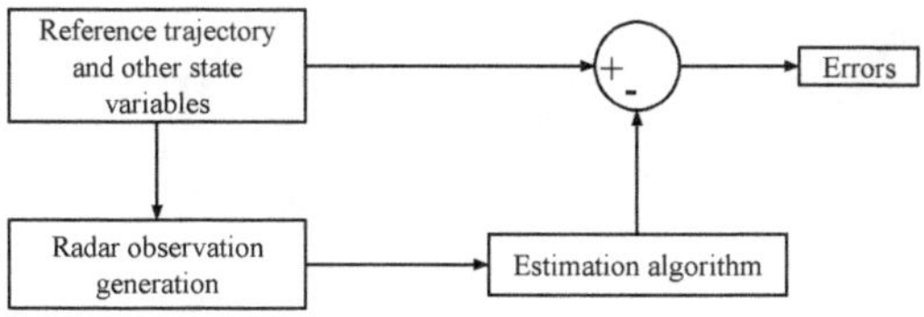

Figure 6.2: Estimation using radar observations.

The radar observations corresponding to the reference trajectory are generated using the measurement equations (6.4) and (6.5). Measurement noise is also included in the observation simulation. We will use these observations in various estimation algorithms to estimate the trajectory of the reentry vehicle and compare the performance. Let us consider the state vector for the reentry vehicle defined in

Chapter 2:

$$\boldsymbol{X} = \begin{bmatrix} x & h & v & \gamma_f & C_D \end{bmatrix}^T \tag{6.6}$$

For all the estimation algorithms, the initial state vector and error covariance are:

$$\widehat{\boldsymbol{X}}(0) = \begin{bmatrix} 101\ km \\ 5\ km \\ 6.050\ km/s \\ -10^\circ \\ 0.7 \end{bmatrix} \tag{6.7}$$

$$\boldsymbol{P}(0) = diag \begin{bmatrix} 6\ km^2 \\ 6\ km^2 \\ .1\ km^2/s^2 \\ .1\ rad^2 \\ .1 \end{bmatrix} \tag{6.8}$$

The process noise covariance matrix for all estimators is selected as

$$\boldsymbol{Q}(t) = 10^{-10} \boldsymbol{I}_{5\times5} \tag{6.9}$$

This particular value of the process noise covariance matrix provides consistent and numerically stable results for multiple runs for all the estimators. Recall that in the Kalman filter framework, measurement Jacobian is required. In this example, the measurement Jacobian is

$$\boldsymbol{H} = \begin{bmatrix} \frac{\partial r}{\partial h} & \frac{\partial r}{\partial x} & 0 & 0 & 0 \\ \frac{\partial E}{\partial h} & \frac{\partial E}{\partial x} & 0 & 0 & 0 \end{bmatrix} \tag{6.10}$$

here

$$\frac{\partial r}{\partial h} = \frac{1}{r}(h + R_E - R\cos\phi) \tag{6.11}$$

$$\frac{\partial r}{\partial x} = -\frac{1}{r}(h + R_E)\sin\phi \tag{6.12}$$

$$\frac{\partial E}{\partial h} = \frac{1}{1+\theta^2}\frac{1-\cos\phi}{\sin\phi} \tag{6.13}$$

$$\frac{\partial E}{\partial x} = \frac{1}{1+\theta^2}\left[-\frac{1}{R_E} + \frac{\theta}{R_E}\cos\phi\right] \tag{6.14}$$

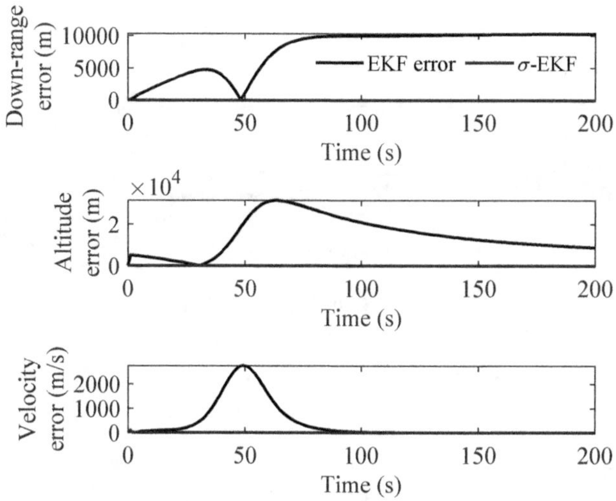

Figure 6.3: Average estimation error of the EKF for 20 runs.

and

$$\theta = \frac{h + R_E - R_E \cos\phi}{R_E \sin\phi} \tag{6.15}$$

The reentry vehicle tracking simulation using the EKF, UKF, SPUKF, and ESPUKF is performed for 20 Monte Carlo runs incorporating different random noise sequences in range and angle observations. The average altitude, down-range and velocity errors, and the standard deviation for the EKF are shown in Figure 6.3.

The mean estimation errors and the standard deviations over the Monte-Carlo runs for the UKF, SPUKF, and ESPUKF are shown separately in Figure 6.4. Table 6.1 provides the mean of the time-averaged state estimation error and standard deviation for various state estimators. Note that the SPUKF and the ESPUKF propagate the state with the accuracy of first and second order Taylor series terms respectively and consequently provides accuracy comparable to the UKF [2]. It can be observed that the estimation error of the EKF is much higher than the UKF and the variants of the UKF. This is because of the highly nonlinear nature of the reentry dynamics.

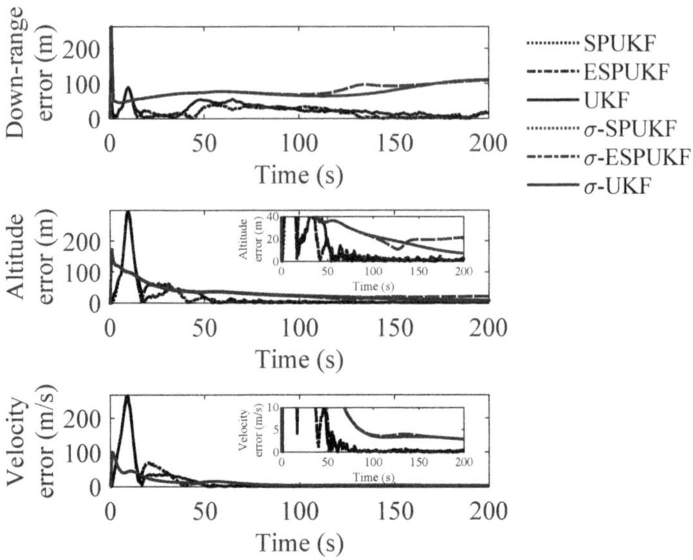

Figure 6.4: Average estimation error of various UKFs for 20 runs.

Table 6.1

Time Averaged Mean Error and Standard Deviation for Various Estimation Algorithms for 20 Runs

	Down-range error (m)		Altitude error (m)		Velocity error (m/s)	
	Mean	σ	Mean	σ	Mean	σ
EKF	7841.541	0.605	14619.623	7.446	376.920	0.825
SPUKF	18.606	80.753	18.120	32.767	15.690	10.643
ESPUKF	18.603	80.845	18.011	32.787	15.390	10.663
UKF	22.924	76.462	16.999	30.759	14.626	10.561

6.3 LAUNCH VEHICLE NAVIGATION USING GPS

Ideally, the position of the launch vehicle should be estimated in three dimensions. However, we will restrict the launch vehicle motion to a 2D plane for simplicity. As a result, the position of the launch vehicle can be expressed as the altitude and downrange distance from the launch site. Other state variables required to express the launch vehicle dynamics are longitudinal speed, flight path angle, and aerodynamic coefficient. In addition to these state variables, GPS receiver clock bias and drift are also included in the state vector.

It should be noted that attitude information is also essential during the launch mission. An INS is used with GPS in practice. A navigation and attitude filter can be designed together or designed separately. We will discuss some aspects of GPS-INS integration in the launch vehicle navigation context later. However, attitude determination algorithms will not be discussed in this chapter as it is out of the book's scope.

6.3.1 Reference Trajectory Generation

Commercial Resupply Service (CRS) -5 mission is selected for the numerical simulation of a launch vehicle. In the CRS-5 mission, a Falcon 9 V1.1 launch vehicle was used. The launch site coordinate is $28.4889°N$, $80.5778°W$. Let us consider the state vector for the launch vehicle defined in Chapter 2:

$$\boldsymbol{X} = \begin{bmatrix} x & h & v_l & \lambda & m & C \end{bmatrix}^T \tag{6.16}$$

For the simulation of the reference trajectory, the initial state vector is

$$\boldsymbol{X}_0 = \begin{bmatrix} 0\,m & 0\,m & 5.6443\,m/s & 1.5708\,rad & 520876\,kg & 0.5 \end{bmatrix}^T \tag{6.17}$$

The mission and launch vehicle-specific parameters for the scenario are provided in Table 6.2 [3]. The initial conditions and the vehicle-specific parameters are used in the launch vehicle dynamics equation discussed in Chapter 2. The reference trajectory is generated by propagating the differential equation using the fourth-order Runge-Kutta numerical integration method with $0.1s$ step size. The process noise standard deviations of $10^{-4}m/s$ and 10^{-3} are considered in acceleration and the time derivative of the aerodynamic coefficient, respectively. The trajectory of the Falcon 9 V1.1 for the specified mission is shown in Figure 6.5a. Figure 6.5b shows the velocity profile of the launch vehicle. At 187s, the first stage separation occurs and a discontinuity in the motion is observed. The flight path angle is plotted in

Table 6.2
Mission and Launch Vehicle Specific Parameters

Mission Parameters	
Payload	2317 kg
Dragon spacecraft mass	4200 kg
Orbit perigee	410 km
Orbit apogee	418 km
Stage 1	
Inert mass	23,100 kg
Propellant mass	395,700 kg
Engine	9× Merlin 1D
Thrust	5886 kN
Specific impulse	282 s
Burnout time	187 s
Stage 2	
Inert mass	3,900 kg
Propellant mass	92,670 kg
Engine	1× Merlin 1D Vac
Thrust	801 kN
Specific impulse	340 s
Burnout time	386 s

Figure 6.6a. At the 50s after launch, a small pitch-over angle is provided to start the gravity turn trajectory. A random walk variation can be observed in the aerodynamic coefficient evolution over time in Figure 6.6b.

Using this reference trajectory, one could generate GPS pseudorange and range-rate observation numerically or use a GNSS simulator to generate the observations and use the data to test the designed navigation filter. For this example with the launch vehicle, the reference trajectory is generated externally using MATLAB. The externally generated trajectory information is converted to the SimGEN (see Section 5.3) compatible user motion command file. This user motion command file is used as the launch vehicle motion input to the SimGEN application. The procedure is shown in Figure 6.7. In this example, the SimGEN directly generated the corresponding pseudorange and carrier range observations with the atmospheric error and receiver clock bias. The observations are used in various estimation algorithms. The estimated states for all the algorithms are compared with the reference

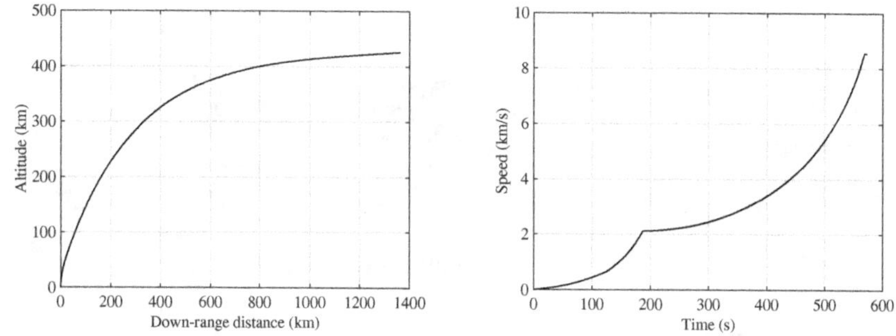

(a) Trajectory **(b)** Velocity profile

Figure 6.5: Falcon 9 V1.1 trajectory and velocity profile.

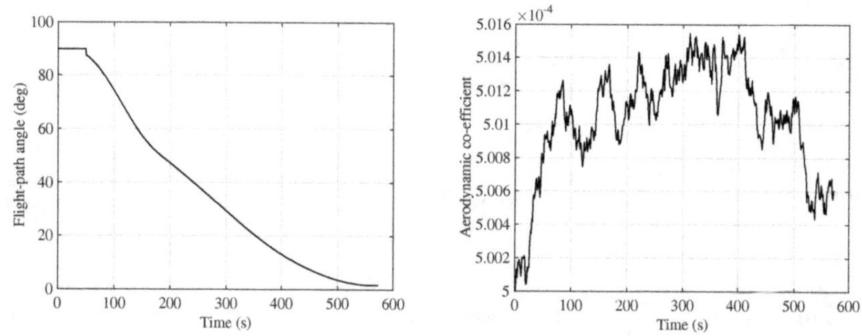

(a) Flight path angle **(b)** Aerodynamic coefficient

Figure 6.6: Falcon 9 V1.1 flight path angle and aerodynamic coefficient.

trajectory for performance comparison of various estimation techniques. The block diagram of this experiment is shown in Figure 6.8.

6.3.2 Implementation of Sequential Estimator

Similar to the GPS-based satellite navigation problem, the receiver clock bias and the drift must be estimated in the navigation filter for the launch vehicle. Essentially the state vector for the navigation filter becomes

$$\boldsymbol{X} = \begin{bmatrix} x & h & v_l & \lambda & m & C & a & b & \dot{b} \end{bmatrix}^T \tag{6.18}$$

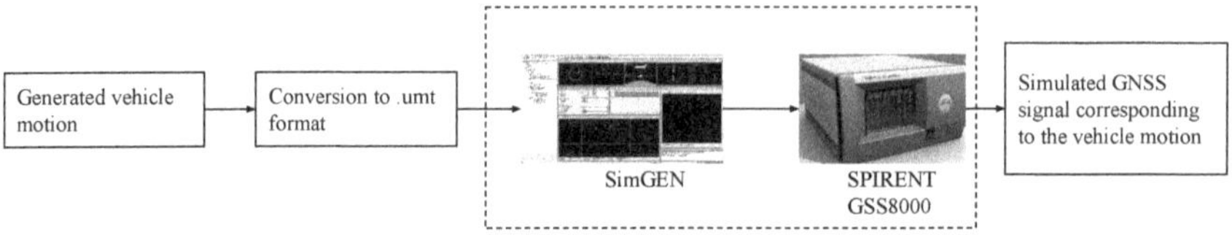

Figure 6.7: Using external trajectory data in SPIRENT simulator.

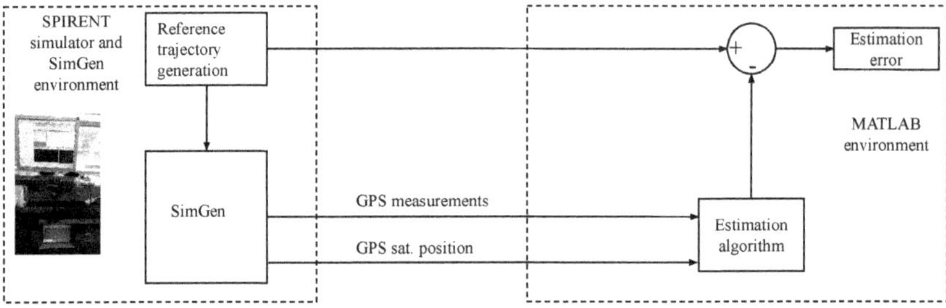

Figure 6.8: GPS observation simulation for a launch vehicle trajectory.

Here x is the down-range distance, h is the altitude, v_l is the longitudinal speed, λ is the flight path angle, m is the mass, C is the aerodynamic coefficient, a is the acceleration, b is the receiver clock bias converted to distance, and $\dot{b}$ is the receiver clock drift converted to speed. The corresponding measurement Jacobian associated with the i-th navigation satellite is

$$\boldsymbol{H_i} = \begin{bmatrix} \frac{\partial \rho_i}{\partial \boldsymbol{r}}\frac{\partial \boldsymbol{r}}{\partial x} & \frac{\partial \rho_i}{\partial \boldsymbol{r}}\frac{\partial \boldsymbol{r}}{\partial h} & 0 & 0 & 0 & 0 & 1 & 0 \\ \frac{\partial \dot{\rho}_i}{\partial \boldsymbol{r}}\frac{\partial \boldsymbol{r}}{\partial x} & \frac{\partial \dot{\rho}_i}{\partial \boldsymbol{r}}\frac{\partial \boldsymbol{r}}{\partial h} & \frac{\partial \dot{\rho}_i}{\partial \boldsymbol{v}}\frac{\partial \boldsymbol{v}}{\partial v_l} & \frac{\partial \dot{\rho}_i}{\partial \boldsymbol{v}}\frac{\partial \boldsymbol{v}}{\partial \lambda} & 0 & 0 & 0 & 1 \end{bmatrix} \tag{6.19}$$

Here, $\boldsymbol{r}$ and $\boldsymbol{v}$ are the launch vehicle position and velocity respectively in the ECEF frame, ρ_i and $\dot{\rho}_i$ are the GPS pseudorange and range rate for the i-th navigation satellite. For multiple satellites, all the associated Jacobians are stacked together row-wise. It should be noted that the computation of the measurement Jacobian in the form shown in (6.19) is done because the GPS satellite position is calculated in the ECEF frame.

The mean state estimation errors and standard deviation in the down-range, altitude, and velocity for the UKF, SPUKF, and the ESPUKF using the pseudorange

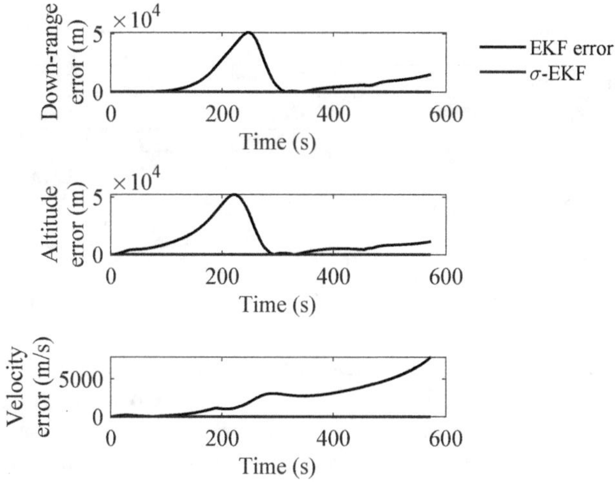

Figure 6.9: Average estimation error of the EKF for 20 runs.

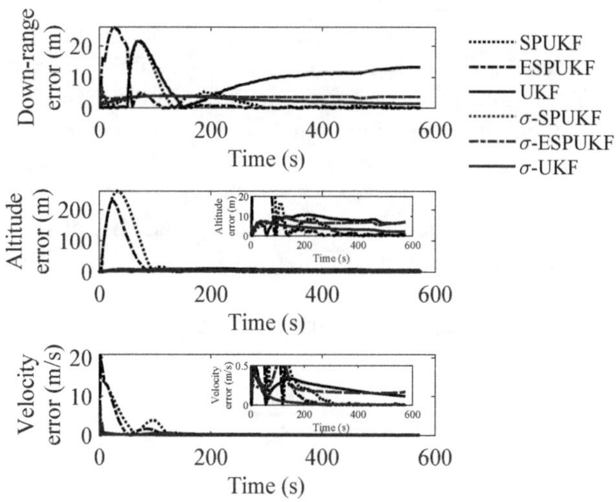

Figure 6.10: Average error and standard deviation for various UKFs for 20 runs.

Table 6.3
Time Averaged Mean Error and Standard Deviation for Various Estimation Algorithms for 20 Runs

	Down-Range Error (m)		Altitude Error (m)		Velocity Error (m/s)	
	Mean	σ	Mean	σ	Mean	σ
EKF	738.017	3.481	783.666	6.083	831.133	0.189
SPUKF	12.229	3.560	11.621	6.782	11.057	0.234
ESPUKF	1.776	3.644	1.602	6.995	1.346	0.253
UKF	14.312	2.542	13.775	4.083	13.367	0.046

and range-rate data for 20 Monte-Carlo runs are shown in Figure 6.10. The estimation performance of the EKF is shown in Figure 6.9. The mean of the time-averaged estimation error and the standard deviation for various estimators are shown in Table 6.3. The UKF and its variants provide better estimation performance in this example compared to the EKF due to the nonlinearity of the launch vehicle dynamics. It is also observed that the ESPUKF provides better performance than the UKF in this estimation problem despite the approximation done in the state and covariance propagation. An explanation of this observation is not available for the time being and it requires further investigation.

Due to the high dynamics of the launch vehicle, it may not be possible for a GPS receiver to acquire GPS signals through all the available channels throughout the trajectory. To examine the performance of various estimation algorithms during the limited availability of the acquired signals, multiple simulations were performed by restricting the number of channels to 4, 6, 8, and 10. The time-averaged PDOP for different numbers of satellites are shown in Figure 6.11. It is observed that, with an increase in the number of channels (i.e., the number of satellites used for navigation), the difference in estimation errors for the EKF and other unscented filters decreases [4].

Recall the standard deviation of the absolute position error σ_r using GNSS for LSE is

$$\sigma_r = PDOP\sigma_\rho \tag{6.20}$$

where PDOP is the position dilution of precision and σ_ρ is the user range error. However, this relation does not hold for the Kalman filter-based state estimation.

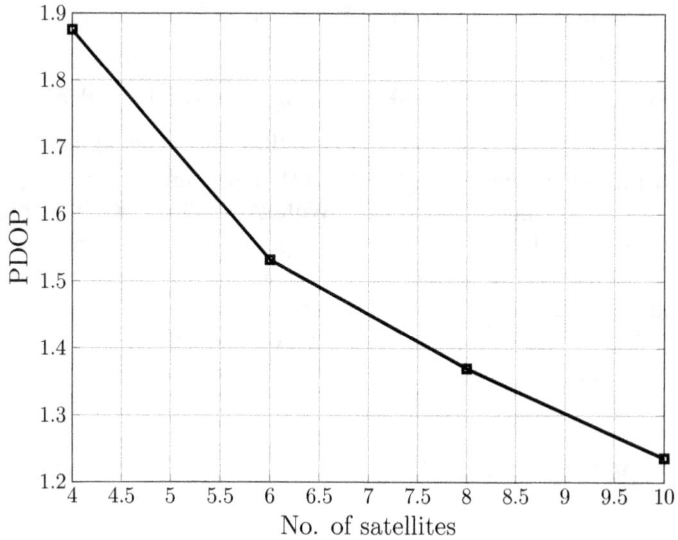

Figure 6.11: Average PDOP variation with number of satellites.

For the EKF it can be shown that [5]

$$D_1^2 = \boldsymbol{C}_{11}\left(\frac{\sigma_x}{\sigma_\rho}\right)^4 + \frac{\sum_{j=2}^{n} \boldsymbol{P}_{1j}^{+}\boldsymbol{C}_{j1}}{\sigma_\rho^2}\left(\frac{\sigma_x}{\sigma_\rho}\right)^2 + \frac{\boldsymbol{OT}_{11}}{\sigma_\rho^4} \tag{6.21}$$

$$D_2^2 = \boldsymbol{C}_{22}\left(\frac{\sigma_y}{\sigma_\rho}\right)^4 + \frac{\sum_{j=3}^{n} \boldsymbol{P}_{2j}^{+}\boldsymbol{C}_{j2}}{\sigma_\rho^2}\left(\frac{\sigma_y}{\sigma_\rho}\right)^2 + \frac{\boldsymbol{OT}_{22}}{\sigma_\rho^4} \tag{6.22}$$

$$D_3^2 = \boldsymbol{C}_{33}\left(\frac{\sigma_z}{\sigma_\rho}\right)^4 + \frac{\sum_{j=3}^{n} \boldsymbol{P}_{3j}^{+}\boldsymbol{C}_{j3}}{\sigma_\rho^2}\left(\frac{\sigma_z}{\sigma_\rho}\right)^2 + \frac{\boldsymbol{OT}_{33}}{\sigma_\rho^4} \tag{6.23}$$

where D_1, D_2, D_3 are the first three diagonal elements of the DOP matrix, $\boldsymbol{C}_{ij}$ is the (i, j)th element of the matrix $\boldsymbol{C} = (\boldsymbol{K}\boldsymbol{K}^T)^-1$, $\boldsymbol{P}_{ij}^{+}$ is the (i, j)th element of the error covariance matrix $\boldsymbol{P}^{+}$, and $\boldsymbol{OT}_{ii}$ are the terms that does not contain the standard deviations in x, y, and z direction - σ_x, σ_y, and σ_z, respectively. These equations imply that the error standard deviation varies hyperbolically with the PDOP. Similar relation can be shown for the UKF as well [6].

For the corresponding average PDOP, considering the magnitude of the position estimation error as ϵ values of average $\left(\frac{\epsilon}{\sigma_\rho}\right)^2$ are plotted in Figure 6.12. It is observed that, for the EKF, SPUKF, ESPUKF, and UKF PDOP vs. average $\left(\frac{\epsilon}{\sigma_R}\right)^2$ curves are hyperbolic in nature. Note that, the position error of the EKF is much higher than the SPUKF, ESPUKF, and UKF for high PDOP values (i.e., when fewer navigation satellites are available). With decreasing PDOP the differences in position errors of the filters reduce. This is expected because the nonlinearity index of the observation $\mathcal{N}_m$ discussed in Chapter 4 increases with increasing PDOP, which is a consequence of the definition of the nonlinearity index. The GDOP ($\sqrt{G_{11}^2 + G_{22}^2 + G_{33}^2 + G_{44}^2}$) can be interpreted as an HS norm of $\boldsymbol{G}$, which is a symmetric matrix, one can write

$$\mathcal{N}_m(t+\delta t, t) \triangleq \sup_{j=1,2\ldots,N} \|\boldsymbol{H}_j(t+\delta t, t) - \overline{\boldsymbol{H}}(t+\delta t, t)\| \times GDOP \tag{6.24}$$

which implies that with the increase in the GDOP (and in turn the PDOP), the measurement nonlinearity increases which impacts the estimator performance.

6.3.2.1 GNSS and INS Integration for Launch Vehicle

In practice, the onboard flight computer on a launch vehicle combines GNSS and INS for navigation. Combining GNSS and INS enables a high-rate navigation solution and makes the navigation algorithm resilient to the brief unavailability of GNSS. As mentioned earlier, one could design a state estimator combining all the state variables of the launch vehicle and the state variable required to describe the error dynamics of the INS. If a multi-GNSS receiver is used, the intersystem clock biases must be included in the state vector.

The error state variables of the INS include bias and bias drifts of accelerometer and gyroscopes, misalignment errors, and scale factors. This is called a tightly coupled estimator. Implementation of a tightly coupled estimator is a complex task as the dimension of the state vector is very large. For example, the dimension of the state vector for expressing only the error dynamics of a strap-down gyroscope is 39 [7]. In addition, the measurement Jacobian must be reformulated accordingly.

Another approach is loosely coupled state estimation. In this approach, the navigation of the launch vehicle using GNSS is performed separately, as described in the previous section, and the error state of the INS is estimated using the solution from the GNSS-based estimation. The error state of the INS is further simplified

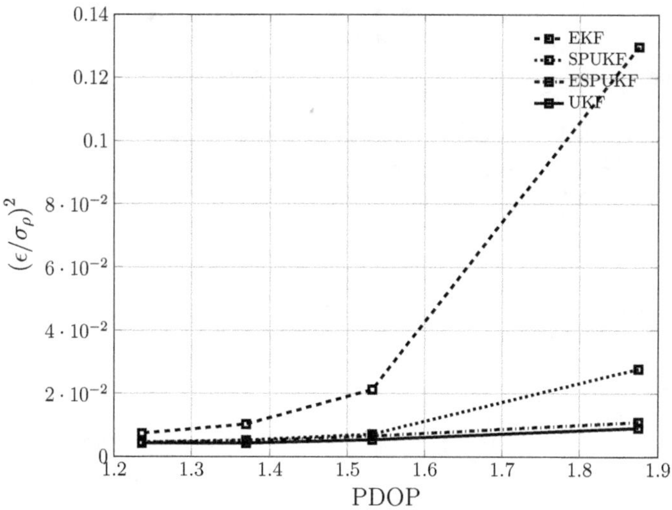

Figure 6.12: $\left(\frac{\epsilon}{\sigma_\rho}^2\right)$ vs. $PDOP$.

based on offline experiments by eliminating some of the state variables that do not have a significant effect on the navigation solution.

6.4 EXAMPLE MATLAB IMPLEMENTATION OF LAUNCH VEHICLE AND REENTRY VEHICLE POSITION ESTIMATION

In this section, we will demonstrate step-by-step MATLAB implementation of a launch vehicle and reentry vehicle position estimation using the UKF. The simulated observations used here are generated based on the simulation method described in Sections 6.2 and 6.3. The results discussed in the previous section can be generated using the codes provided in this section.

6.4.1 Reentry Vehicle Tracking

The MATLAB code for the reentry vehicle tracking using the ground-based range, azimuth, and elevation observations is given below. The code file is presented in several snippets here. The first snippet is the initialization of the variables and loading of the data. The data is loaded from the file `Reentry_Tdata.mat` that contains the true trajectory and the measurements. The initial state and the corresponding initial standard deviation of the state variables are loaded from the file `RV_INI.mat`.

```
clc
%clear all
%% Reentry vehicle position estimation program
load Rentry_Tdata % true trajectory and measurements
load RV_INI
global m R_e M Qk t_step R
m = 4000;
R_e = 6371*10^3; %Radius of earth in m
M = 10*10^3; % down-range of the radar
t_step = 0.1;
Qk = process_cov;
R = noise_cov;
X_true = Rentry_Tdata.Tstate;
Z = Rentry_Tdata.measurement;
```

The next snippet is the initialization of the state vector, the covariance matrix, and the measurement noise covariance matrix, and creates the `model` structure described in Section 4.7 which contains the function handle for the reentry vehicle dynamics, measurement model, and the Jacobians.

```
STATE = struct([]);

STATE(1).X = X_INI;
STATE(1).STD = STD_INI;
STATE(1).P = zeros(5);
%STATE(1).K = 0;
for i = 1:length(X_INI)
 STATE(1).P(i,i) = STD_INI(i)^2;
end
X_est(1,:) = X_INI;
std(1,:) = STD_INI;
RVMODEL.dynamics = @rentrydyn_UKF;
```

```
RVMODEL.msmtf = @h_rentry;
RVMODEL.Fjacobian = @rentryjacobian;
RVMODEL.Mjacobian = @rentry_mjacobian;
time = TIME;
RSD = [];
```

The next code snippet is the main loop of the UKF. The UKF function is called for each time step and the estimated state and the error covariance are stored in the `STATE` structure. `RV_UKFdata` stores all the necessary data for further analysis.

```
fprintf('Estimating position')
profile on
for i = 1:length(time)-1
 msmt.Z = Z(:,i+1);
 msmt.dt =1;
 [statec , rsd] = UKF(STATE(i),msmt,RVMODEL);
 STATE(i+1).X = statec.X;
 STATE(i+1).P = statec.P;
 STATE(i+1).STD = statec.STD;
 RSD = [RSD rsd];
 X_est(i+1,:) = statec.X';
 std(i+1,:) = statec.STD;
 prcsd = i/time(end)*100;
 clc
 fprintf('%.0f percent processed\n',prcsd)
end
PR = profile('info');
profile off
error = X_true - X_est;
RV_UKFdata.X = X_est;
RV_UKFdata.STD = std;
RV_UKFdata.error = error;
RV_UKFdata.profile = PR;
save RV_UKFdata RV_UKFdata
```

6.4.2 GNSS-based Navigation of Launch Vehicle

The MATLAB code for the GNSS-based navigation of the launch vehicle using the UKF is given below. The code file is presented here through multiple code blocks. The first block shown below loads the observation, reference trajectory, initial state, and covariance information. `LV_data.mat` contains the GPS pseudorange and range rate observations and `Flight_data.mat` conatains the reference trajectory considered as the truth. The initial state and the corresponding initial standard deviation of the state variables are loaded from the file `LV_INI.mat`. This block also initializes the process and measurement noise covariance matrices.

```
load LV_data.mat
load LV_INI.mat
load FLT_data.mat
load Flight_data.mat
% Global variables
global R_E T g pitch_ctl me_dot dT STG_m_dot dD D R Qk t_step GPSPVT phi lambda Az R_LS b NS

R_E = 6378137; %m
b = 6356752.314; %m
rho = 1.225; %kg/m^3
g = 9.81; %m/s
sigma_R = LV_INI.R(1); %m
sigma_RR = LV_INI.R(2);
%RB = [sigma_R^2 0; 0 sigma_RR^2];
Qk = LV_INI.Q;
phi= lat;
lambda = lon;
Az = beta;
R_LS = R0;
NS = LV_INI.NS;
ref_LLA = [lat lon 0];
```

The next snippet is for the launch vehicle-specific constants. It contains information regarding a number of stages, including stage masses, thrust, specific impulse, and burnout time. It also declares the flags for the stage separation, pitch control input, and the time at which the pitch control is initiated.

```
%% Vehicle-specific constants
I_SP1 = 282; %s
I_SP2 = 340; %s
d = 3.66; %m
S1_burnoutT = 187;
S2_burnoutT = 386;
T1 = 5886000;%N
T2 = 801000;%N
STG1_PM = 395700; %kg
STG2_PM = 92670; %kg
stg1_M = 23100; %kg
stg2_M = 3900; %kg
Pay_load = 2370 + 4200; %kg
M = STG1_PM + STG2_PM+stg1_M + stg2_M + Pay_load;
A = pi*d^2/4; %frontal area
 %% Flags
STG_flag1 = 0;
STG_flag2 = 0;
FPA_flag = 0;
ATM_flag = 0;

%% Control sequence
Pitch_ctl_alt = 138;%m
pitch_rate = -deg2rad(0.005); %rad/s
```

The next code snippet is for the model handles and estimator initialization. The model handles are used by the UKF function to propagate the state and measurement models. The state model is given by `lvdyn_UKF` and the measurement model is given by `GPS_msmt`. The Jacobians of the state and measurement models are given by `LV_Fjacobian` and `GPS_Mjacobian`, respectively. Note that the Jacobians are not required for the UKF implementation. For the EKF, SPUKF, or ESPUKF, the Jacobians are required.

```
%% model handles
```

```
LV_model.dynamics = @lvdyn_UKF;
LV_model.Fjacobian = @LV_Fjacobian;
LV_model.msmtf = @GPS_msmt;
LV_model.Mjacobian = @GPS_Mjacobian;
%% Initialization
M0 = M-T1/(I_SP1*g)*0.04;
v0 = I_SP1*g*log(M/M0);
D = rho*v0^2*A*0.5;
t_step = 0.1;
X0 = [0 0 v0 pi/2 M0 0.5 (T1-D)/M0-g 300 0]' + LV_INI.delX';
TIME = 0:1:S1_burnoutT + S2_burnoutT;

STATE = struct([]);
STATE(1).P = zeros(9);
STATE(1).X = X0;
STATE(1).STD = LV_INI.STD;
for i = 1:9
 STATE(1).P(i,i) = STATE(1).STD(i)^2;
end
%% EKF
RSDR = [];
RSDV = [];
```

The next code block implements the UKF. Lines 4 to 62 implement the pitch control, stage separation, and atmospheric drag computation logic. The rest of the lines are for the UKF implementation and storing the state, error covariance, and standard deviations.

```
1 %fprintf('Estimating position')
2 profile on
3 for i = 1:length(TIME)- 1
4  %% Pitch-over
5  if STATE(i).X(2) >=Pitch_ctl_alt
6  if FPA_flag ==0
7   pitch_ctl = pitch_rate;
```

```
 %STATE(i).X = LV_data.X(i,:)';
 FPA_flag = 1;
else
 pitch_ctl = 0;
end
else
pitch_ctl = 0;
end
%% Stage 1
if STATE(i).X(5) <= M - STG1_PM
 if STG_flag1 ==0
 STATE(i).X(5) = STATE(i).X(5) - stg1_M;
 STG_m_dot = - stg1_M;
 STG_flag1 = 1;
 dT = T2-T1;
 else
 STG_m_dot = 0;
 dT = 0;
 end
 T = T2;
 me_dot = T2/(I_SP2*g);
else
 T = T1;
 me_dot = T1/(I_SP1*g);
 STG_m_dot = 0;
 dT = 0;
end
%% Stage 2
if STATE(i).X(5) <= M - STG1_PM - stg1_M - STG2_PM
if STG_flag2 ==0
 STATE(i).X(5) = STATE(i).X(5) - stg2_M;
 STG_m_dot = - stg2_M;
 dT = -T2;
 STG_flag2 = 1;
else
```

```
    STG_m_dot = 0;
    dT = 0;
  end
  T = 0;
  me_dot = 0;
  end
  %% Atmospheric drag
  if STATE(i).X(2)>= 30000
    if ATM_flag == 0
      rho = 0;
      dD = -D;
      D = 0.5*rho*STATE(i).X(3)^2*A*STATE(i).X(6);
      ATM_flag = 1;
    else
      dD = 0;
    end
  else
    dD = 0;
    D = 0.5*rho*STATE(i).X(3)^2*A*STATE(i).X(6);
  end
  MSMT.PRN = GPSM(i).PRN(1:NS);
  MSMT.Z = [GPSM(i).R(1:NS);GPSM(i).RR(1:NS)];
  MSMT.dt = GPSM(i).dt;
  %% Constructing R
  R = zeros(2*NS);
  for k = 1:2*NS
    if k<= NS
      R(k,k) = sigma_R^2;
    else
      R(k,k) = sigma_RR^2;
    end
  end
  GPSPVT = PVT_GPS(i);
  [state_C, rsd] = UKF(STATE(i),MSMT,LV_model);
  STATE(i+1).X = state_C.X;
```

```
  STATE(i+1).P = state_C.P;
  STATE(i+1).STD = state_C.STD;
  RSDR = [RSDR; rsd(1:NS)];
  RSDV = [RSDV; rsd(NS+1:2*NS)];
end
```

The last code block is for storing the processing time, converting the position and velocity state vector from the local frame to the ECEF frame, and storing the data in a structure. In particular, the `LV_UKF_data` structure contains estimation error, standard deviation, position, and velocity in the ECEF frame.

```
PR = profile('info');
ex_time = PR.FunctionTable(3).TotalTime/PR.FunctionTable(3).NumCalls;
msmtupdt_time = PR.FunctionTable(14).TotalTime/PR.FunctionTable(14).NumCalls...
 + PR.FunctionTable(15).TotalTime/PR.FunctionTable(15).NumCalls;
for i = 1:length(STATE)
 EST.X_est(i,:) = STATE(i).X';
 EST.STD(i,:) = STATE(i).STD';
end

for i = 1:length(EST.X_est)
  [R_ECEF,V_ECEF] = Co_ord_change(EST.X_est(i,:));
 LV_UKF_data.R_ECEF(i,:) = R_ECEF;
 LV_UKF_data.V_ECEF(i,:) = V_ECEF;
end
error = LV_data.X(1:length(EST.X_est),:) - EST.X_est(:,1:7);
LV_UKF_data.ER = Ref_R_ECEF(1:length(LV_UKF_data.R_ECEF),:) - LV_UKF_data.R_ECEF;
LV_UKF_data.EV = Ref_V_ECEF(1:length(LV_UKF_data.V_ECEF),:) - LV_UKF_data.V_ECEF;
LV_UKF_data.error = error;
LV_UKF_data.TIME = TIME;
```

```
LV_UKF_data.ExTime = ex_time;
LV_UKF_data.MsmtUpdtT = msmtupdt_time;
save('LV_UKF_data.mat','LV_UKF_data')
```

6.5 PRACTICAL CONSIDERATIONS

In practice, designing and implementing a navigation algorithm for reentry or launch missions requires extensive numerical simulation, tuning the designed algorithm from previous data or offline experiments and hardware-in-loop simulation for verification and validation. One of the main challenges in the hardware implementation of navigation filters is model and process noise mismatch. In the simulation, the reference trajectory is generated using a numerical model of the system. Based on this reference trajectory, the performance of the navigation algorithm is evaluated, and tuning is performed. However, the simulated scenario does not always capture all the aspects of the real-world system, and the process noise may not match the actual launch or reentry vehicle. A model and process noise mismatched navigation algorithm can perform poorly on the physical system. A model mismatch is often handled by parameter estimation, where some of the model parameters are estimated online rather than selecting the parameters used in the simulation scenario. The process noise covariance is estimated from previous mission data and offline experiments.

One important aspect of estimator design is tuning (i.e., selecting the process noise covariance matrix). This is an iterative process where one should start with an initial guess of the process noise covariance and perform Monte Carlo simulations and adjust the covariance until the estimator output becomes consistent (i.e., the estimation error and the residual are zero mean and the residual is white noise [8]). In addition, it is also necessary to evaluate the sensitivity of the designed estimator with respect to various factors such as initialization, tuning of the process noise covariance, and measurement rate. The sensitivity of the analysis with respect to these factors is done by observing the variation of a cost function, which is necessarily a function of the estimation error and the standard deviation by varying the above-mentioned factors [9]. It should be noted that these factors are not exhaustive and depend on the mission.

An estimator for navigation or tracking should be less susceptible to spurious measurement errors which may appear from time to time due to various reasons. This is done by rejecting measurements with residuals higher than a predefined threshold. Another approach is performing $\mathcal{X}^2$ test to identify outliers in the measurement residues.

Turning now to some practicalities of launch vehicle navigation using GNSS observation, a regular onboard GNSS receiver will not be able to track the GNSS signal due to a very high Doppler rate due to the high dynamics of the launch vehicle. As a result, a specialized GNSS receiver capable of acquiring GNSS signals at a high acceleration must be used. While GNSS is used for onboard navigation of launch vehicles, the use of the same is challenging in the atmospheric reentry mission. Due to plasma deposition during the reentry through the atmosphere, GNSS signals cannot be acquired. An additional mechanism must be implemented in the navigation filter to prevent significant estimation errors during the unavailability of the GNSS signals. High-accuracy INS can restrict the INS error and estimate the INS error, considering the vehicle dynamics are modeled accurately. This will help reduce the time to first fix (TTFF) of the GNSS receiver and help in the fast reacquisition of GNSS signals.

Hardware implementation of a navigation algorithm must be done judiciously as computational resources and power are limited onboard launch or reentry vehicles. A navigation engineer must choose a suitable estimation algorithm based on the available computational resource on the flight computer and the required navigation solution rate. The choice of estimator also depends on the nonlinearity of the vehicle dynamics, as discussed in Chapter 4. The UKF is more suitable for the launch and reentry mission as both systems are highly nonlinear. However, the computational complexity of the UKF is high. The simulation results in the previous sections show that the ESPUKF provides performance similar to the UKF. Since the processing time of the ESPUKF is almost 80% less than the UKF, this algorithm is suitable for the launch and reentry missions.

References

[1] G. R. Curry, *Radar System Performance Modeling*. Boston: Artech House, 2005.

[2] S. K. Biswas, L. Qiao, and A. G. Dempster, "A Novel a Priori State Computation Strategy for the Unscented Kalman Filter to Improve Computational Efficiency," *IEEE Transactions on Automatic Control*, vol. 62, no. 4, pp. 1852–1864, Apr. 2017.

[3] "CRS-5 Fifth Commercial Resupply Services Flight to the International Space Station," NASA, Tech. Rep., 2014.

[4] S. K. Biswas, L. Qiao, and A. G. Dempster, "Computationally Efficient Unscented Kalman Filtering Techniques for Launch Vehicle Navigation using a Space-Borne GPS Receiver," *ION GNSS+*, Portland, Oregon, USA, 2016, p. 9.

[5] S. K. Biswas, "Computationally Efficient Non-Linear Kalman Filters for On-Board Space Vehicle Navigation," Ph.D. dissertation, UNSW Sydney, Sydney, Australia, 2017.

[6] S. K. Biswas, L. Qiao, and A. G. Dempster, "Effect of PDOP on performance of Kalman Filters for GNSS-based space vehicle position estimation," *GPS Solutions*, vol. 21, no. 3, pp. 1379–1387, Jul. 2017.

[7] J. Farrell, *Aided Navigation: GPS With High Rate Sensors* (Electronic Engineering). New York: McGraw-Hill, 2008.

[8] Y. Bar-Shalom, X. R. Li, and T. Kirubarajan, *Estimation With Applications to Tracking and Navigation: Theory Algorithms and Software*. John Wiley & Sons, 2004.

[9] M. Rhudy, Y. Gu, J. Gross, S. Gururajan, and M. R. Napolitano, "Sensitivity Analysis of Extended and Unscented Kalman Filters for Attitude Estimation," *Journal of Aerospace Information Systems*, vol. 10, no. 3, pp. 131–143, Mar. 2013.

Chapter 7

Navigation and Tracking in Lunar and Deep Space Missions

7.1 OVERVIEW

Human activities in space are not confined to earth-bound orbits or near-earth space. Since the beginning of space exploration, various nations have independently or jointly planned and executed various lunar and deep space missions. These activities not only include sending probes to other planets in the solar system, and their natural satellites, landing rovers on Mars and the Moon, but also redirecting asteroids from their orbit. While the motivations behind these missions have been debated from time to time across the community, there is no doubt that the scientific and technological outcome of these missions is enormous. For example, we have already discussed that our celebrated navigation algorithm the EKF was developed for a lunar landing and now it has found its way to our handheld mobile devices, not only for personal navigation applications but also for other sensor fusions.

Having discussed the extent of human space exploration in deep space, now let us focus on the navigation of spacecraft during lunar or any other deep space missions. From a mathematical perspective, our problem statement does not change for these missions. Similar to other types of space vehicle navigation discussed in previous chapters, the goal here remains the same: find the best possible estimate of spacecraft position and velocity, given the uncertainty in the spacecraft dynamics and a set of noisy measurements. From the dynamics provided in Chapter 2, one can easily conclude that the deep space navigation problem is also a nonlinear estimation problem and can be solved using a sequential estimator (for example, the EKF). However, deep space navigation comes with some practical challenges due to the large distance spacecraft have to travel, sparse observations, and unavailability of very precise onboard clocks. While some of the above-mentioned problems can be handled by modifying the navigation algorithm, some need better technology. In this chapter, we will focus on the algorithm segment only.

The rest of the chapter is organized as follows: first, we will elaborate on the challenges in deep space or lunar navigation compared to the navigation of satellites, a launch vehicle, or a reentry vehicle. Subsequently, we will discuss the navigation filter design for deep space missions, possible observations, formulation of the state transition matrix, and the measurement Jacobian. Afterwards, we will show some numerical results for the navigation of spacecraft in lunar transfer trajectory. We will conclude the chapter with some recent developments in deep space navigation and future directions.

7.2 CHALLENGES IN LUNAR AND DEEP SPACE NAVIGATION

Currently, the observations for lunar and deep space missions rely on ground-based atomic clocks. The signal round-trip time from a ground station is used as range observation. Since the signal has to travel a large distance in deep space and a lunar mission, it creates a significant delay (i.e. the observation received at any given instance is essentially a past observation). Accounting for these delays in the navigation algorithm is extremely important for lunar and deep space missions.

One of the major challenges in the onboard navigation in lunar and deep space missions is the lack of onboard precise clock. One could argue that the atomic clocks that are used in GNSS can also be used in a deep space mission. However, it should be noted that the atomic clocks in GNSS are not stable in the long term and needs frequent updates from the ground. As a result, the long-term stability of GNSS atomic clocks does not satisfy the requirement of deep space missions as the frequent update from the ground station is not possible and communication suffers significant delay as the distance of the spacecraft increases from the earth.

The above-mentioned difficulties differentiate navigation in lunar and deep space missions from near-earth missions. In this chapter, we will focus on ground-based tracking of a spacecraft in the lunar transfer trajectory as an example. The discussed method can be easily extended to other interplanetary missions.

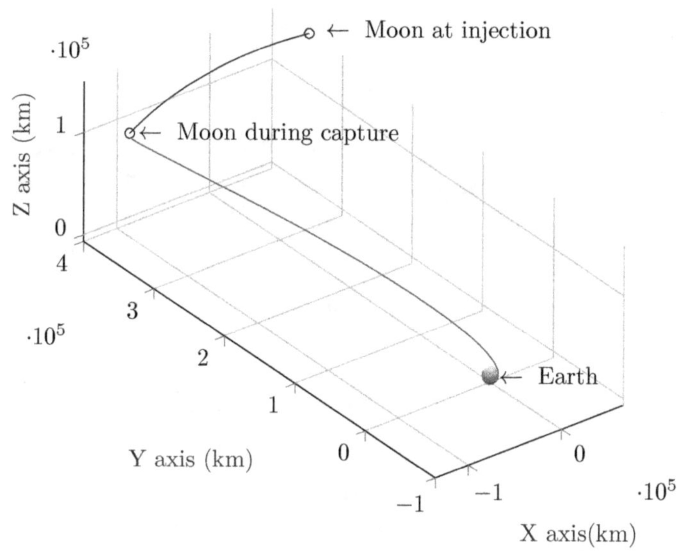

Figure 7.1: A reference lunar transfer trajectory.

7.3 GROUND-BASED TRACKING OF SPACECRAFT IN LUNAR TRANSFER TRAJECTORY

In this numerical simulation example, we consider a spacecraft in the lunar transfer trajectory that is being tracked by four ground stations: Goldstone Space Communications Complex, Madrid Space Communications Complex, Canberra Space Communications Complex, and Indian Deep Space Network Byalalu based on the visibility of the spacecraft from each ground station. The observations are azimuth, elevation, range, and range rate.

The reference trajectory, which is considered the true trajectory, is generated considering solar gravitational perturbation, solar radiation pressure, lunar gravitational perturbation, and J2 effect of the earth and the Moon. The Earth and Moon's gravitational coefficients are obtained from Joint Gravity Model-3 (JGM-3) [1] and LP150Q [2], respectively. The reference trajectory of the spacecraft is shown in Figure 7.1. The initial time is 0.00 hours, September 20, 2011 UTC, and the duration of the simulation is $3,347$ minutes. The same trajectory is shown in Figure 7.2 in the baricentric rotating frame.

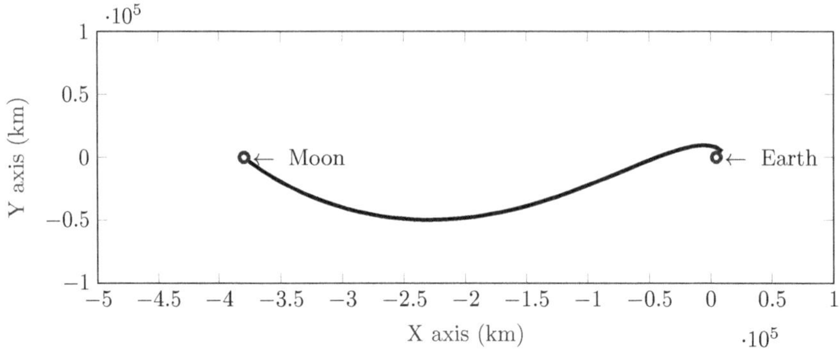

Figure 7.2: A lunar transfer trajectory in the barycentric rotating frame.

Table 7.1
Observation Noise Standard Deviations

Observation	σ
Range	10 m
Range rate	0.1 cm/s
Azimuth	0.01°
Elevation	0.01°

Using the reference trajectory and ground station coordinates, the observations are simulated. Numerically generated noises are incorporated as well to simulate the uncertainty in the observations. The noise standard deviations used in the simulation are provided in Table 7.1. For designing an EKF-based navigation program, we need to obtain the expression of the state transition matrix, the measurement Jacobian, and the delays associated with the measurements, which are discussed in the subsequent parts of the section.

7.3.1 State Transition Matrix

The partial derivative of acceleration with respect to the state vector must be calculated to compute the state transition matrix $\mathbf{\Phi}$. The acceleration comprises the acceleration due to the Earth as a point mass, acceleration due to the gravitational perturbations, and acceleration due to other celestial bodies. Partial derivatives for each of the terms can be calculated separately and added. The partial derivative of the acceleration considering the Earth as a point mass will be

$$\frac{\partial \ddot{\boldsymbol{r}}_p}{\partial \boldsymbol{r}} = -GM\left(\frac{1}{r^3}\boldsymbol{I}_{3\times3} - 3\boldsymbol{r}\frac{\boldsymbol{r}^T}{r^5}\right) \tag{7.1}$$

The partial derivative of the acceleration term due to the largest gravitational coefficient C_{20}; that is, for J2 term in the ECEF frame can be written as [3]

$$\frac{\partial \ddot{x}_{20}}{\partial x} = \frac{GM}{R_e^3}\frac{1}{2}\left[C_{20}V_{4,2} - \frac{4!}{2!}C_{20}V_{4,0}\right] \tag{7.2}$$

$$\frac{\partial \ddot{x}_{20}}{\partial y} = \frac{GM}{R_e^3}\frac{1}{2}C_{20}W_{4,2} \tag{7.3}$$

$$\frac{\partial \ddot{x}_{20}}{\partial z} = \frac{GM}{R_e^3}3C_{20}V_{4,1} \tag{7.4}$$

$$\frac{\partial \ddot{y}_{20}}{\partial z} = \frac{GM}{R_e^3}3C_{20}W_{4,1} \tag{7.5}$$

$$\frac{\partial \ddot{z}_{20}}{\partial z} = \frac{GM}{R_e^3}\frac{4!}{2!}C_{20}V_{4,0} \tag{7.6}$$

where R_e is the radius of the Earth and

$$W_{n,m} = \frac{R_e}{r}^{n+1} P_{nm}\, sin\phi\, sin(m\lambda)$$

$$V_{n,m} = \frac{R_e}{r}^{n+1} P_{nm}\, sin\phi\, cos(m\lambda)$$

where P_{nm} is the associate Legendre polynomial of degree n and order m, and ϕ and λ are the latitude and longitude of the spacecraft, respectively. In the ECI frame, the partial derivative can be expressed as

$$\left(\frac{\partial \ddot{\boldsymbol{r}}_{J2}}{\partial \boldsymbol{r}}\right)_{ECI} = \boldsymbol{R}^{-1}(t)\left(\frac{\partial \ddot{\boldsymbol{r}}_{J2}}{\partial \boldsymbol{r}}\right)_{ECEF} \boldsymbol{R}(t) \tag{7.7}$$

Here $\boldsymbol{R}(t)$ is the ECEF to ECI transformation matrix.

The partial derivative of the acceleration due to i-th celestial body of mass M_i at $\boldsymbol{s_i}$ position in the ECI frame is given by

$$\frac{\partial \ddot{\boldsymbol{r}}_i}{\partial \boldsymbol{r}} = -GM_i\left(\frac{1}{|\boldsymbol{r}-\boldsymbol{s}_i|^3}I_{3\times3} - 3(\boldsymbol{r}-\boldsymbol{s}_i)\frac{(\boldsymbol{r}-\boldsymbol{s}_i)^T}{|\boldsymbol{r}-\boldsymbol{s}_i|^5}\right) \tag{7.8}$$

Then the partial derivative of the total acceleration considering the above-mentioned perturbations and k celestial objects can be written as

$$\frac{\partial \ddot{\boldsymbol{r}}}{\partial \boldsymbol{r}} = \frac{\partial \ddot{\boldsymbol{r}}_p}{\partial \boldsymbol{r}} + \frac{\partial \ddot{\boldsymbol{r}}_{J2}}{\partial \boldsymbol{r}} + \sum_{i=1}^{k} \frac{\partial \ddot{\boldsymbol{r}}_i}{\partial \boldsymbol{r}} \tag{7.9}$$

7.3.2 Partial Derivative of Measurements

The expression of partial derivatives of measured range ρ, azimuth A, and elevation E with respect to position vector $\boldsymbol{r}$ were already provided in Section 5.3.2. In addition to the above observations, we will use range rate and hence the partial derivative of the range rate with respect to the position and velocity vector are also required in the EKF. The partial derivative of the range rate with respect to $\boldsymbol{r}$ is

$$\frac{\partial \dot{\rho}}{\partial \boldsymbol{r}} = \frac{r_{gs}\dot{\boldsymbol{r}}_{gs}^T - \dot{r}_{gs}\boldsymbol{r}_{gs}^T}{r_{gs}^2} \boldsymbol{R}_{TF}\boldsymbol{R}_{FI} \tag{7.10}$$

The partial derivative of the range rate with respect to the velocity vector is

$$\frac{\partial \dot{\rho}}{\partial \boldsymbol{v}} = \frac{\boldsymbol{r}_{gs}{}^T}{r_{gs}} \boldsymbol{R}_{TF}\boldsymbol{R}_{FI} \tag{7.11}$$

where $\boldsymbol{r}_{gs}$ is the position vector of the satellite from the ground station in the topocentric frame, $\boldsymbol{R}_{TF}$ is the rotation matrix from the ECEF frame to the topocentric frame, and $\boldsymbol{R}_{FI}$ is the rotation matrix from the ECI frame to the ECEF frame.

7.4 DELAY IN OBSERVATION

The ground-based range measurement in deep space missions is generally performed using two-way ranging to ensure clock bias-free range observation. In the two-way ranging technique, a signal is transmitted from the ground station to the spacecraft. The spacecraft retransmits the signal to the ground station. The ground station determines the round trip time τ of the signal. The round trip time consists of up-link travel time τ_u and down-link travel time τ_d. The down-link signal travel time for a ground station with position vector $\boldsymbol{r}_g(t)$ is given by [3]

$$\tau_d = \frac{1}{c}|\boldsymbol{r}(t - \tau_d) - \boldsymbol{r}_g(t)| \tag{7.12}$$

The up-link signal travel time is

$$\tau_u = \frac{1}{c}|\boldsymbol{r}(t-\tau_d) - \boldsymbol{r}_g(t-\tau_d-\tau_u)| \tag{7.13}$$

Considering $\tau = \tau_u + \tau_d$ the two-way range measurement ρ observed at time t can be modeled as

$$\begin{aligned}\rho &= \frac{1}{2}(\rho(t-\tau)+\rho(t))\\ &= \frac{c}{2}(\tau_u+\tau_d)\end{aligned} \tag{7.14}$$

Then one can approximate the following

$$\rho\left(t-\frac{\tau}{2}\right) \approx \frac{1}{2}(\rho_u+\rho_d) \tag{7.15}$$

This approximation implies the average two-way range at time t is nearly equal to the geometric range between the spacecraft and the ground station at time $\left(t-\frac{\tau}{2}\right)$. From this observation, one can conclude that the range measurement is delayed by $\frac{\tau}{2}$ second.

Two-way Doppler measurements are obtained by integrating Doppler counts over a time interval t_i. Let the range observations be recorded at time t_1 and t_2, and $t_t = t_2 - t_1$. Considering the uplink and downlink delays corresponding to epoch t_1 and t_2 as τ_{1u}, τ_{1d} and τ_{2u}, τ_{2d}, respectively, the average range rate measurement can be modeled as

$$\bar{\dot{\rho}}(t) = \frac{c}{2}\frac{(\tau_{2u}+\tau_{2d})-(\tau_{1u}+\tau_{1d})}{t_c} \tag{7.16}$$

It is shown [4] that the average range rate at time t_2 can be approximated as

$$\begin{aligned}\bar{\dot{\rho}}(t_2) &\approx \dot{\rho}\left(t_1+\frac{t_i}{2}-\frac{1}{2}(\tau_{2d}+\tau_{1d})\right)\\ &= \dot{\rho}\left(t_2-\frac{1}{2}(t_i+\tau_{2d}+\tau_{1d})\right)\end{aligned} \tag{7.17}$$

This equation shows that the Doppler measurement is delayed by $\frac{1}{2}(t_i+\tau_{2d}+\tau_{1d})$ second.

From the previous discussion, it is evident that both range and range rate observation have different delays. Note that when the spacecraft is in the near-earth space these delays can be neglected. However, as the distance of the spacecraft increases these delays also increase and can become of the order of 2 seconds near the Moon. As a result the Kalman filter for this case needs to be designed carefully to handle the delays.

7.5 IMPLEMENTATION OF NAVIGATION ALGORITHM

Having discussed the state transition matrix and delays in observations, let us now discuss the implementation of the navigation filter for tracking the spacecraft in the lunar transfer trajectory shown in Figure 7.1. Since the time delay $\tau_{r_{n+1}}$ for the azimuth A, elevation E, range σ observations is different from the time delay $\tau_{rr_{n+1}}$ for the range-rate observation $\dot{\rho}$ at epoch t_{n+1}, two updates must be performed for a set of $A, E, \rho, \dot{\rho}$ observations at that epoch.

The sequence of state updates for corresponding delayed observations is shown in Figure 7.3. The range rate delay is larger than the range delay and hence first the EKF update for the range rate must be done, which results in the a posteriori state vector $\widehat{\boldsymbol{X}}^{+}(t_{n+1} - \tau_{rr_{n+1}})$ as shown in Figure 7.3. This updated state is then propagated to the epoch $t_{n+1} - \tau_{r_{n+1}}$ corresponding to the range, azimuth, and elevation observation. The propagated state is then updated using these observations which results in $\widehat{\boldsymbol{X}}^{+}(t_{n+1} - \tau_{r_{n+1}})$ as shown in Figure 7.3.

The implementation of the EKF for handling the measurement delays is shown in Figure 7.4. Since two measurement updates are involved at two different time-varying delays, the entire estimator can be represented as two consecutive estimators from a modular implementation perspective: estimator 1 — for range rate update and estimator 2 — for azimuth, elevation, and range update. The estimated state at the epoch t_n epoch is associated with the time $t_n - \tau_{r_n}$. At epoch t_{n+1} the range rate is associated with the epoch $t_{n+1} - \tau_{rr_{n+1}}$, the A, E, ρ is associated with the time $t_{n+1} - \tau_{r_{n+1}}$. The state is propagated to $t_{n+1} - \tau_{rr_{n+1}}$ in the estimator 1 and then the a posteriori state is computed using the range-rate corresponding to the epoch $t_{n+1} - \tau_{rr_{n+1}}$. The updated state is the input to the estimator 2, where the state is propagated to $t_{n+1} - \tau_{r_{n+1}}$ and then the a posteriori state is computed using A, E, and ρ observation.

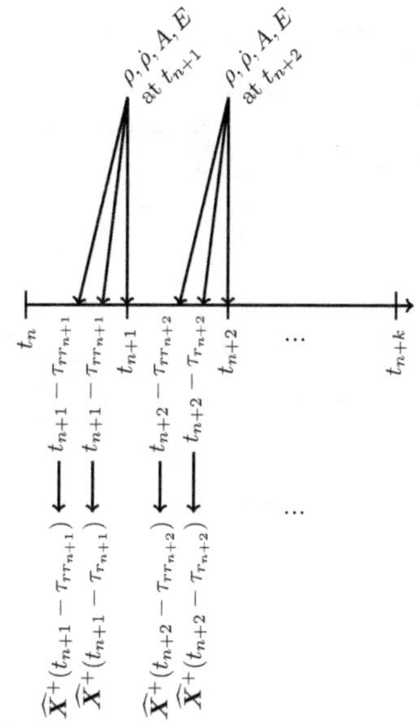

Figure 7.3: Observation delays and the corresponding state update sequence.

The initial state vector for the above estimator is

$$\boldsymbol{X}(0) = \begin{bmatrix} \boldsymbol{r}(0) \\ \boldsymbol{v}(0) \end{bmatrix} = \begin{bmatrix} 3648.4\ km \\ -5170.3\ km \\ -1797.8\ km \\ 9.0717\ km/s \\ 5.3546\ km/s \\ 3.0410\ km/s \end{bmatrix} \tag{7.18}$$

and the corresponding error covariance matrix is

$$\boldsymbol{P}(0) = diag\left(\begin{bmatrix} 9\ km^2 \\ 9\ km^2 \\ 9\ km^2 \\ 0.0001\ km^2/s^2 \\ 0.0001\ km^2/s^2 \\ 0.0001\ km^2/s^2 \end{bmatrix}\right) \tag{7.19}$$

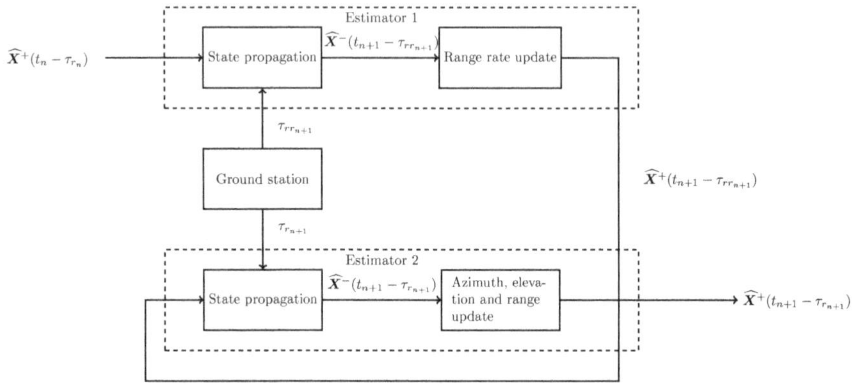

Figure 7.4: The EKF for observations with delays.

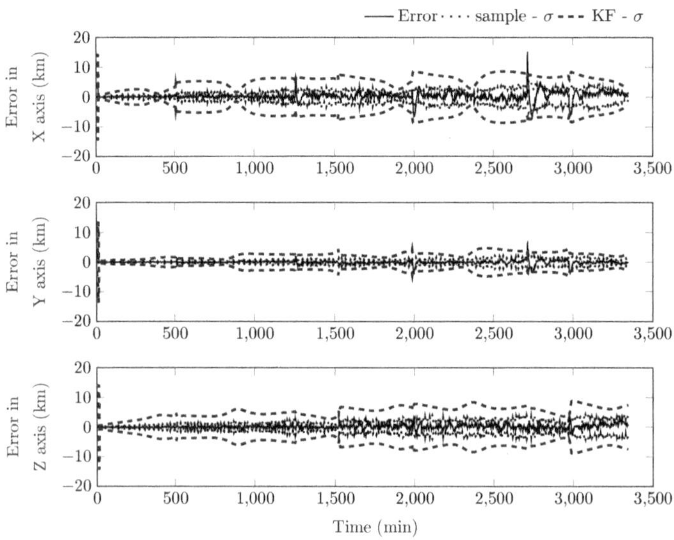

Figure 7.5: Position estimation performance for 20 Monte Carlo simulations.

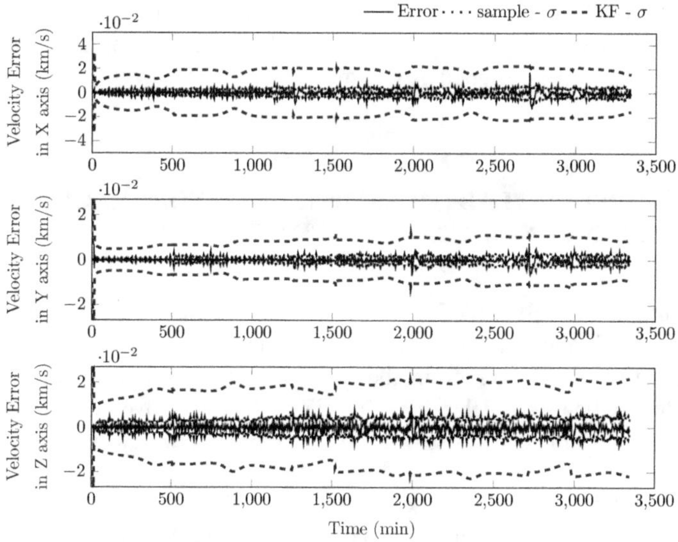

Figure 7.6: Velocity estimation performance.

The initial epoch is 0.0 hours, September 20, 2011 UTC. Note that the continuous-time process noise $\boldsymbol{Q}(t)$ is defined as

$$\boldsymbol{Q}(t) = \begin{bmatrix} \eta_v(t)\boldsymbol{I}_{3\times3} & \boldsymbol{0}_{3\times3} \\ \boldsymbol{0}_{3\times3} & \eta_a(t)\boldsymbol{I}_{3\times3} \end{bmatrix} \tag{7.20}$$

where η_v and η_a are the velocity and acceleration power spectral density, respectively. The discrete process noise $\boldsymbol{Q}_k$ for timestep Δt will be

$$\boldsymbol{Q}_k = \begin{bmatrix} (\eta_v\Delta t + \eta_a\frac{\Delta t^3}{3})\boldsymbol{I}_{3\times3} & \eta_a\frac{\Delta t^2}{2}\boldsymbol{I}_{3\times3} \\ \eta_a\frac{\Delta t^2}{2}\boldsymbol{I}_{3\times3} & \eta_a\Delta t\boldsymbol{I}_{3\times3} \end{bmatrix} \tag{7.21}$$

η_v and η_a are selected as $10^{-8}(m/s)^2/Hz$ and $10^{-6}(m/s^2)^2/Hz$ in the EKF through tuning. The measurement rate is 0.1 Hz for each ground station. Figures 7.5 and 7.6 show the average position and velocity estimation errors for 20 Monte Carlo simulations, the sample standard deviation, and the mean of standard deviations over the 20 Monte Carlo simulations. The results show that the sample standard deviation is less than the estimator standard deviation, and the average error for 20 simulations lies within the estimated standard deviation most of the time indicating the consistency of the designed estimator.

7.6 EXAMPLE MATLAB IMPLEMENTATION FOR TRACKING A SPACECRAFT IN LUNAR TRANSFER TRAJECTORY

In this section, we present the MATLAB implementation of ground-based tracking of a spacecraft in the scenario described in the previous sections. This EKF-based implementation is different from the implementations presented in the previous chapter due to the use of delay-handling EKF. Here, we have not divided the code file into several snippets because of a long for-loop. Let us now inspect the code file step by step.

The first line loads the range, azimuth, elevation, and range rate observations and incorporates measurement noise. The next line initializes the initial state vector, error covariance, process noise, and measurement noise covariances. The for-loop runs the delay handling EKF sequentially. Line 5 selects the observation from the ground stations based on the elevation angle. Lines 7 to 18 propagate the state and the error covariance for the next 10 seconds.

Note that we need to propagate the state and the error covariance until the epoch considering the range rate delay. Lines 30 through 36 adjust the state propagation until the time corresponding to the range rate delay. Line 39 updates the state vector and the error covariance using the EKF.

Lines 56 to 60 propagate the range rate updated state and error covariance till the epoch corresponding to the range delay. Line 63 updates the state vector and the error covariance using the EKF.

Lines 73 to 91 propagate the state and the error covariance until the 10-second interval, whereas lines 95 to 115 propagates the state and the error covariance until the 10-second interval when range, azimuth, and elevation measurements are not available. Lines 118 to 126 update the state and the error covariance using the propagated state and error covariance when no observation is available.

```
1  run measurement_processing
2  run initialisation
3  wB = waitbar(0,'Progress');
4  for i = 1:iteration
```

```
    obs_info = meaurement_selection(i, IDSN, GS, MD, CB);

    % Propagation
    t_sim = (i-1)*sampt:1:i*sampt;
    t_step = 1;
    [X_prop, cov, P, temp_var1] =
        state_covariance_propagation(R_sc3_ini,...
         R_v3_ini,R_m3_ini,R_mv3_ini,t_sim,t_step, P, '
        full');
    P1 = P;
    covariance((i-1)*sampt + 1:i*sampt-2,:) = cov;
    X_filter((i-1)*sampt+1:i*sampt,:) = X_prop(2:end,:);

    R_sc3_ini = temp_var1.R_sc3_ini;
    R_v3_ini = temp_var1.R_v3_ini;
    R_m3_ini = temp_var1.R_m3_ini;
    R_mv3_ini = temp_var1.R_mv3_ini;

    if obs_info.Elevation > 5*pi/180
        phi = obs_info.phi;
        lambda = obs_info.lambda;
        %% Initialization for next 2s
        C = strcmp(obs_info.flag,flag_prev);
        if C < 1
            count = 0;
        end
        count = count + 1;

        t_sim = i*sampt-2: (2 - obs_info.RR_delay)/5 :i*
        sampt - obs_info.RR_delay;
        t_step = (2 - obs_info.RR_delay)/5;

        [X_prop, cov, P, temp_var2] =
        state_covariance_propagation(R_sc3_ini,...
```

```
            R_v3_ini,R_m3_ini,R_mv3_ini,t_sim,t_step, P,
        'mupdate');
        X_priori_RR = X_prop(end,:);
        time_RR = i*sampt - obs_info.RR_delay;

        [State_covariance, X_update, delRR, del_X,
        KALMAN_GAIN, RC_RR] = RR_UPDATE(X_priori_RR,
        time_RR,obs_info.phi...
            ,obs_info.lambda,obs_info.Range_Rate,P,R_rr);


        state_cov_rr(i,:) = [State_covariance(1,1)
        State_covariance(2,2) State_covariance(3,3)...
                              State_covariance(4,4)
        State_covariance(5,5) State_covariance(6,6)];

        INNOVATION_RR(i) = delRR;
        K_RR(i,:) = KALMAN_GAIN;
        del_XRR(i,:) = del_X';
        recovRR(i) = RC_RR;
        R_sc3_ini = X_update(1:3);
        R_v3_ini = X_update(4:6);
        R_m3_ini = temp_var2.R_m3_ini;
        R_mv3_ini = temp_var2.R_mv3_ini;

        if count == 10 && C == 1
            % Propagating till RAE measurement
            Z = [obs_info.Range obs_info.Azimuth obs_info
        .Elevation];
            t_sim = time_RR:(obs_info.RR_delay - obs_info
        .range_delay)/5:i*sampt - obs_info.range_delay;
            t_step = (obs_info.RR_delay - obs_info.
        range_delay)/5;
            [X_prop, cov, P, temp_var3] =
        state_covariance_propagation(R_sc3_ini,...
```

```
            R_v3_ini,R_m3_ini,R_mv3_ini,t_sim,t_step, P, 'mupdate');
        X_priori_RAE = X_prop(end,:);
        time_RAE = i*sampt - obs_info.range_delay;
        [State_covariance_RAE, X_update, delRAE, del_X, KALMAN_GAIN_RAE, RC_RAE] = RAE_UPDATE(X_priori_RAE,i*sampt,phi,...
                lambda,Z,P,R_rae);
        state_cov_r(i,:) = [State_covariance_RAE(1,1) State_covariance_RAE(2,2) State_covariance_RAE(3,3)...
                    State_covariance_RAE(4,4) State_covariance_RAE(5,5) State_covariance_RAE(6,6)];
        INNOVATION_RAE(i,:) = delRAE';
        K_RAE(:,:,i) = KALMAN_GAIN_RAE;
        del_XR(i,:) = del_X(:,1)';
        recovRAE(i,:) = [RC_RAE(1,1) RC_RAE(2,2) RC_RAE(3,3)];
        R_sc3_ini = X_update(1:3);
        R_v3_ini =  X_update(4:6);
        %% Propagation till end of 10s
        t_sim = i*sampt - obs_info.range_delay:obs_info.range_delay/5:i*sampt;
        t_step = obs_info.range_delay/5;
        [X_prop, cov, P, temp_var4] = state_covariance_propagation(R_sc3_ini,...
            R_v3_ini,R_m3_ini,R_mv3_ini,t_sim,t_step, P, 'mupdate');
        X_priori = X_prop(end,:);
        state_cov(i,:) = [P(1,1) P(2,2) P(3,3)...
                        P(4,4) P(5,5) P(6,6)];
        state(i+1,:) = X_priori;
        R_sc3_ini = X_priori(1:3);
        R_v3_ini = X_priori(4:6);
```

```
        R_m3_ini = temp_var4.R_m3_ini;
        R_mv3_ini = temp_var4.R_mv3_ini;

        t_step = 1;
        [cov,~,~] = propagate2(temp_var1.Rm,temp_var1
    .Xp,P1,t_step,i*sampt-2:1:i*sampt);
        covariance(i*sampt-1:i*sampt,:) = cov;

        X_filter(i*sampt,:) = state(i+1,:);
        covariance(i*sampt,:) = state_cov(i,:);
        count = 0;
    else
        %% Propagation till end of 10s
        t_sim = i*sampt - obs_info.RR_delay:obs_info.
    RR_delay/5:i*sampt;
        t_step = obs_info.RR_delay/5;
        [X_prop, cov, P, temp_var5] =
    state_covariance_propagation(R_sc3_ini,...
            R_v3_ini,R_m3_ini,R_mv3_ini,t_sim,t_step,
     P, 'mupdate');
        X_priori = X_prop(end,:);
        state_cov(i,:) = [P(1,1) P(2,2) P(3,3)...
                          P(4,4) P(5,5) P(6,6)];
        state(i+1,:) = X_priori;

        R_sc3_ini = X_priori(1:3);
        R_v3_ini = X_priori(4:6);
        R_m3_ini = temp_var5.R_m3_ini;
        R_mv3_ini = temp_var5.R_mv3_ini;

        t_step = 1;
        [cov,~,~] = propagate2(temp_var1.Rm,temp_var1
    .Xp,P1,t_step,i*sampt-2:1:i*sampt);
        covariance(i*sampt-1:i*sampt,:) = cov;

```

```
                X_filter(i*sampt,:) = state(i+1,:);
                covariance(i*sampt,:) = state_cov(i,:);
                flag_prev = obs_info.flag;
            end
        else
            [cov,P,~] = propagate2(temp_var1.Rm,temp_var1.Xp,P,t_step,i*sampt-2:1:i*sampt);
            covariance(i*sampt-1:i*sampt,:) = cov;
            state(i+1,:) = X_prop(end,:);
            state_cov(i,:) = [P(1,1) P(2,2) P(3,3) P(4,4) P(5,5) P(6,6)];

            R_sc3_ini = temp_var1.R_sc3_ini1;
            R_v3_ini = temp_var1.R_v3_ini1;
            R_m3_ini = temp_var1.R_m3_ini1;
            R_mv3_ini = temp_var1.R_mv3_ini1;
        end
        waitbar(i/iteration,wB,sprintf('Simulation %3.2f %% completed',...
            (i/(iteration))*100));
    end
    delete(wB);
    X_filter = [R0 V0;X_filter];
    cov_EKF_WD = covariance;
    error_EKF = X_true(1:length(X_filter),:) - X_filter;
```

7.7 SUMMARY AND FUTURE DIRECTIONS

In this chapter, we have discussed ground-based tracking in lunar and deep space missions. The state transition matrix, measurement Jacobian matrix formulation, and delays associated with observations were presented. An EKF design for addressing the observation delays is discussed. Finally, a numerical simulation example of tracking a spacecraft in a lunar transfer trajectory was demonstrated.

As discussed in the previous sections, the current deep space navigation and tracking technology rely heavily on ground stations, due to the unavailability of onboard clocks with the required precision and scarcity of accurate observations other than the ground-based azimuth, elevation, range, and range rate.

To address these major hurdles, recently NASA has demonstrated a trapped-ion atomic clock in space that provides long-term stability and drift performance up to an order of magnitude compared to current space clocks [5]. This will enable one-way range and range rate measurement and, consequently, onboard navigation will be possible in deep space.

However, onboard deep space navigation without ground station support has also been demonstrated in NASA's Double Asteroid Redirection Test (DART) [6] mission. The final phase of the mission relied on autonomous vision-based navigation, guidance, and control of the DART spacecraft to identify Dimorphos from the image, maintain a collision course, and impact it in real-time. This mission demonstrated fully autonomous navigation capability in a deep space mission for the first time. It is expected that in the future optical, radiometric, and other available observations will be fused using suitable sequential estimators to navigate in deep space.

References

[1] B. D. Tapley *et al.*, "The joint gravity model 3," *Journal of Geophysical Research: Solid Earth*, vol. 101, no. B12, pp. 28 029–28 049, 1996.

[2] R. B. Roncoli, "Lunar constants and models document," *JPL D-32296, Sept*, 2005.

[3] O. Montenbruck and E. Gill, *Satellite Orbits: Models, Methods, and Applications*, 1st ed. Springer-Verlag, 2002.

[4] B Kruger, "The range rate error due to the averaging techniques of doppler measurements," Tech. Rep., 1965.

[5] E. A. Burt *et al.*, "Demonstration of a trapped-ion atomic clock in space," *Nature*, vol. 595, no. 7865, pp. 43–47, Jul. 2021.

[6] A. S. Rivkin *et al.*, "The Double Asteroid Redirection Test (DART): Planetary Defense Investigations and Requirements," *The Planetary Science Journal*, vol. 2, no. 5, p. 173, Oct. 2021.

Appendix

This appendix provides some frequently used algorithms used in satellite navigation and orbit determination. The first algorithm focuses on computing the position and velocity vectors $\boldsymbol{r}$ and $\boldsymbol{v}$ of a satellite in the ECI frame from Keplerian elements, while the second algorithm deals with converting the ECI position and velocity vectors back to Keplerian elements. The third algorithm is for solving the Lambert problem (i.e., velocity computation using two position vectors). These algorithms are often used in space vehicle navigation, tracking, and orbit determination during the initialization of the estimator. The first two algorithms use true, eccentric, and mean anomalies of a satellite in an elliptical orbit. We will first discuss these anomalies and then continue to the algorithms.

A.1 ANOMALIES

Recall the Keplerian elements are semi-major axis a, eccentricity e, inclination i, right ascension of the ascending node Ω, argument of perigee ω, and mean anomaly M. These two algorithms also require two additional parameters in the intermediate stages — true anomaly ν and eccentric anomaly E. These anomaly parameters are defined as follows:

- *True anomaly* ν is the angle between the perigee and the satellite's current position vector in the orbital plane relative to the focus of the ellipse.
- *Eccentric anomaly* E is the angle between the perigee and the satellite's current position vector projected on the auxiliary circle relative to the center of the ellipse.
- *Mean anomaly* M represents the angular distance covered by the object along its orbit since passing through the periapsis at a constant rate. If the orbital

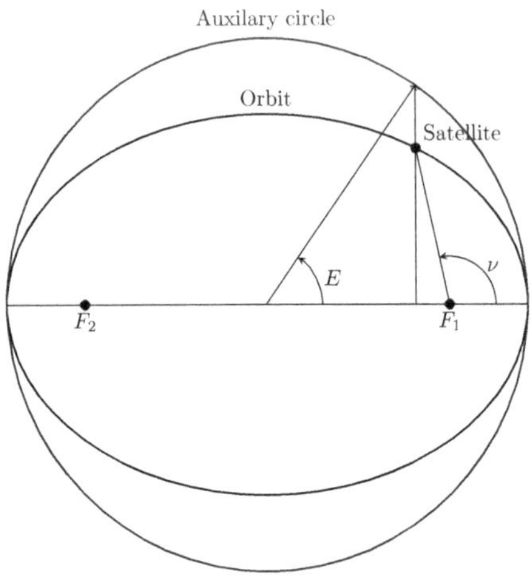

Figure A.1: True and eccentric anomalies in an elliptical orbit.

period is T, then the mean anomaly M at time t is given by $M = \frac{2\pi}{T}(t - \tau)$, where τ is the time of periapsis passage.

The relation between M and E is given by Kepler's equation:

$$M = E - e \sin E \tag{A.1}$$

The relation between E and ν is given by:

$$\nu = 2 \arctan \left(\sqrt{\frac{1+e}{1-e}} \tan \frac{E}{2} \right) \tag{A.2}$$

The true and eccentric anomalies are shown in Figure A.1.

A.2 KEPLERIAN ELEMENTS TO ECI POSITION AND VELOCITY

Here we consider all the angles are in radian. The algorithm is as follows:

Algorithm A.1 Keplerian Elements to ECI Position and Velocity

1: Compute E by solving Kepler's equation numerically
2: Compute ν from E using (A.2)
3: Compute the distance r from the central body to the satellite in the orbital plane using $r = a(1 - e\cos E)$
4: Compute the position vector $\boldsymbol{r}_o = [x_o\ y_o\ z_o]^T$ in the orbital plane: $[x_o\ y_o\ z_o]^T = r[\cos\nu\ \sin\nu\ 0]$
5: Compute the velocity vector in the orbital plane: $\boldsymbol{v}_o = \frac{\sqrt{\mu a}}{r}[-\sin E\ \sqrt{1-e^2}\cos E\ 0]^T$
6: Compute the transformation matrix $\boldsymbol{R}_o^{ECI}$ from the orbital plane to the ECI frame using ω, Ω and i: $\boldsymbol{R}_o^{ECI} = \boldsymbol{R}_Z(-\Omega)\boldsymbol{R}_x(-i)\boldsymbol{R}_z(-\omega)$
7: $\boldsymbol{r} = \boldsymbol{R}_o^{ECI}\boldsymbol{r}_o$ and $\boldsymbol{v} = \boldsymbol{R}_o^{ECI}\boldsymbol{v}_o$

A.3 ECI POSITION AND VELOCITY TO KEPLERIAN ELEMENTS

Algorithm A.2 ECI Position and Velocity to Keplerian Elements

1: Compute the angular momentum vector $\boldsymbol{h} = [h_x \; h_y \; h_z]^T = \boldsymbol{r} \times \boldsymbol{v}$

2: Compute i using $i = \cos^{-1} \frac{h_x}{h}$

3: Compute the vector $\boldsymbol{n} = [n_x \; n_y \; n_z]^T$ in the direction of the ascending node using $n = [0\;0\;1]^T \times \boldsymbol{h}$

4: Compute eccentricity vector $\mathbf{e}$ using $\mathbf{e} = \frac{\boldsymbol{v} \times \boldsymbol{h}}{\mu} - \frac{\boldsymbol{r}}{r}$

5: Compute ν using

$$\nu = \begin{cases} \cos^{-1} \frac{\mathbf{e}^T \boldsymbol{r}}{er} & \text{if } \boldsymbol{r}^T \boldsymbol{v} \geq 0 \\ 2\pi - \cos^{-1} \frac{\mathbf{e}^T \boldsymbol{r}}{er} & \text{otherwise} \end{cases}$$

6: Compute E using $E = 2 \tan^{-1} \frac{\tan \frac{\nu}{2}}{\sqrt{\frac{1+e}{1-e}}}$

7: Compute Ω using

$$\Omega = \begin{cases} \cos^{-1} \frac{n_x}{n} & \text{if } n_y \geq 0 \\ 2\pi - \cos^{-1} \frac{n_x}{n} & \text{if } n_y < 0 \end{cases}$$

8: Compute ω using

$$\omega = \begin{cases} \cos^{-1} \frac{\boldsymbol{n}^T \mathbf{e}}{ne} & \text{if } e_z \geq 0 \\ 2\pi - \cos^{-1} \frac{\boldsymbol{n}^T \mathbf{e}}{ne} & \text{if } e_z < 0 \end{cases}$$

9: Compute M using Kepler's equation

10: Compute a using $a = \frac{r}{1 - e \cos E}$

A.4 LAMBERT PROBLEM

Algorithm A.3 Velocity Computation by Solving the Lambert Problem

---

Calculate magnitudes of the given position vectors $\mathbf{r}_1$ and $\mathbf{r}_2$ and transfer angle $\Delta\theta$

$$\Delta\theta = \cos^{-1}\frac{\mathbf{r}_1 \cdot \mathbf{r}_2}{r_1 r_2}$$

Define $l = \frac{r_1+r_2}{4\sqrt{r_1 r_2}\cos\frac{\Delta\theta}{2}} - \frac{1}{2}$ and $m = \frac{\mu\Delta t^2}{\left\{2\sqrt{r_1 r_2}\cos\frac{\Delta\theta}{2}\right\}^3}$

Initialize $y = 1$

while value of y changes **do**

  $x_1 = \frac{m}{y^2} - l$

  $x_2 = \frac{4}{3}\left(1 + \frac{6x_1}{5} + \frac{6\cdot 8x_1^2}{5\cdot 7} + \frac{6\cdot 8\cdot 10x_1^3}{5\cdot 7\cdot 9} + ...\right)$

  $y = 1 + x_2(l + x_1)$

end while

if elliptical orbit **then**

  $\cos\frac{\Delta E}{2} = 1 - 2x_1$

  $p = \frac{r_1 r_2(1-\cos\Delta\theta)}{r_1+r_2-2\sqrt{r_1 r_2}\cos\frac{\Delta\theta}{2}\cos\frac{\Delta E}{2}}$

else if hyperbolic orbit **then**

  $\cosh\frac{\Delta H}{2} = 1 - 2x_1$

  $p = \frac{r_1 r_2(1-\cos\Delta\theta)}{r_1+r_2-2\sqrt{r_1 r_2}\cos\frac{\Delta\theta}{2}\cosh\frac{\Delta H}{2}}$

end if

Calculate the Lagrange coefficients $f = 1 - \frac{r_2}{p}(1-\cos\Delta\theta)$ and $g = \frac{r_1 r_2 \sin\Delta\theta}{\sqrt{\mu p}}$

Calculate derivatives $\dot{f} = \sqrt{\frac{\mu}{p}}\left(\frac{1-\cos\Delta\theta}{p} - \frac{1}{r_1} - \frac{1}{r_2}\right)\tan\frac{\Delta\theta}{2}$
and $\dot{g} = 1 - \frac{r_1}{p}(1 - \cos\Delta\theta)$

Calculate velocities $\mathbf{v}_1 = \frac{1}{g}(\mathbf{r}_2 - f\mathbf{r}_1)$ and $\mathbf{v}_2 = \dot{f}\mathbf{r}_1 + \frac{\dot{g}}{g}(\mathbf{r}_2 - f\mathbf{r}_1)$

Nomenclature

ACRONYMS

CRLB	Cramer-Rao lower bound
DOP	dilution of precision
DORIS	doppler orbitography and radio positioning integrated by satellite
DSN	deep space network
ECEF	earth-centered earth-fixed frame
ECI	earth centered inertial frame
EKF	extended Kalman filter
ESP-PF	extrapolated single propagation particle filter
ESPT	extrapolated single propagation technique
ESPUKF	extrapolated single propagation unscented Kalman filter
FPE	Fokker-Planck equation
GDOP	geometric dilution of precision
GEO	geosynchronous orbit
GLONASS	globalnaya navigatsionnaya sputnikovaya sistema
GNSS	global navigation satellite system
GPS	global positioning system
GRAPHIC	group and phase ionospheric correction
HEO	high earth orbit
HS norm	Hilbert-Schmidt norm

IGS	international GNSS services
INS	inertial navigation system
LEO	low earth orbit
LSE	least square estimation
MEO	medium earth orbit
NavIC	navigation using Indian constellation
PDOP	position dilution of precision
QZSS	quasi-Zenith satellite system
RAAN	right ascension angle of the ascending node
RNSS	regional navigation satellite system
SLR	satellite laser ranging
SPUKF	single propagation unscented Kalman filter
SSUKF	spherical simplex unscented Kalman filter
SSUT	spherical simplex unscented transform
UKF	unscented Kalman filter
UT	unscented transform

SYMBOLS

$\widehat{(\cdot)}^{+}$	a posteriori
$\widehat{(\cdot)}^{-}$	a priori
a	semi-major axis
β	elevation angle
C	aerodynamic coefficient
c	speed of light

D	aerodynamic drag
e	eccentricity
ϵ_E	Hilbert-Schmidt norm of EKF error
ϵ_U	Hilbert-Schmidt norm of UKF error
$\widehat{(\cdot)}$	estimated variable
ϵ_Φ	random noise in carrier range measurement
ϵ_ρ	random noise in pseudorange measurement
$\boldsymbol{f}$	nonlinear system function
g	gravitational acceleration
γ_f	flight path angle
$\boldsymbol{H}$	Jacobian of $\boldsymbol{h}$
h_a	altitude
$\boldsymbol{h}$	nonlinear measurement function
h	number of integration steps
I	ionospheric error
i	inclination
J_n	zonal hermonics
$\boldsymbol{\mathcal{J}}$	Jacobian of $\boldsymbol{f}$
$\boldsymbol{K}$	Kalman gain
k	discrete time index
$\dot{m}_e$	exhaust mass flow rate
m	mass
μ_e	gravitational parameter of the Earth
n	number of state variables in the state vector
$\mathcal{N}_m$	nonlinearity index of measurement model
$\mathcal{N}_s$	nonlinearity index of system model
$\boldsymbol{\nu}$	process noise vector

$\boldsymbol{\omega}$	measurement noise vector
$\boldsymbol{\Phi}$	state transition matrix
$\boldsymbol{P}$	error covariance matrix
Φ_i	carrier range from the user vehicle to the navigation satellite i
$\boldsymbol{Q}$	process noise covariance matrix
$\boldsymbol{R}$	measurement noise covariance matrix
ρ_i	pseudorange from the user vehicle to the navigation satellite i
r	range
$\boldsymbol{r}$	position vector
R_E	local radius of the Earth
ρ	pseudorange
σ	standard deviation
T	tropospheric error
θ	azimuth angle
t	time
ν	true anomaly
v	speed
$\boldsymbol{v}$	velocity vector
Ω	right ascension angle of the ascending node
ω	argument of perigee
x	down-range
$\boldsymbol{X}$	state vector
$\boldsymbol{Z}$	measurement vector

About the Authors

SANAT K. BISWAS

Sanat K. Biswas is an assistant professor with IIIT Delhi. He received his BE degree from Jadavpur University in 2010, his MTech degree in aerospace engineering from IIT Bombay in 2012, and his PhD degree in computationally efficient unscented Kalman filters for space vehicle navigation from the University of New South Wales (UNSW), Sydney, in 2017.

At IIIT Delhi he leads the Space Systems Laboratory and is involved in developing algorithms for space situational awareness, NavIC reflectometry receiver for remote sensing applications and precise point positioning (PPP) of low earth orbit Satellites. Dr. Biswas serves on the technical committee on Space Communications and Navigation (SCAN), and the technical committee on Space Traffic Management (STM) of the International Astronautical Federation. He was the recipient of the 2014 Emerging Space Leaders Grant from the International Astronautical Federation, 2019 Early Career Research Award from the Department of Science and Technology, India, Young Scientist Award 2020 and 2021, from the International Union of Radio Science (URSI), and 2020 Harry Rowe Mimno Award from the IEEE Aerospace and Electronic Systems Society.

ANDREW G. DEMPSTER

Andrew G. Dempster received his BE and MEngSc degrees from UNSW Australia, in 1984 and 1992, respectively, and his PhD degree in efficient circuits for signal processing arithmetic from the University of Cambridge, Cambridge, U.K., in 1995.

He is director of the Australian Centre for Space Engineering Research (ACSER), UNSW Australia. He was a system engineer and project manager for the first GPS receiver developed in Australia in the late 1980s and has been involved in satellite navigation ever since. He has published in the areas of arithmetic circuits, signal processing, biomedical image processing, satellite navigation, and space systems. His current research interests are in satellite navigation receiver design and signal processing, areas where he has six patents, new location technologies, and space systems.

Index

Artech House
GNSS Technology and Applications Library

Elliott Kaplan and Christopher Hegarty, Series Editors

A-GPS: Assisted GPS, GNSS, and SBAS, Frank van Diggelen

All Source Positioning, Navigation, and Timing, Rongsheng (Ken) Li

Applied Satellite Navigation Using GPS, GALILEO, and Augmentation Systems, Ramjee Prasad and Marina Ruggieri

Digital Terrain Modeling: Acquisition, Manipulation, and Applications, Naser El-Sheimy, Caterina Valeo, and Ayman Habib

Geographical Information Systems Demystified, Stephen R. Galati

GNSS Applications and Methods, Scott Gleason and Demoz Gebre-Egziabher

GNSS Interference Threats and Countermeasures, Fabio Dovis, editor

GNSS Markets and Applications, Len Jacobson

GNSS Receivers for Weak Signals, Nesreen I. Ziedan

GNSS for Vehicle Control, David M. Bevly and Stewart Cobb

GPS/GNSS Antennas, B. Rama Rao, W. Kunysz, R. Fante, and K. McDonald

Implementing e-Navigation, John Erik Hagen

Inertial Navigation Systems Analysis, Kenneth Britting

Introduction to GPS: The Global Positioning System, Second Edition, Ahmed El-Rabbany

Location-Based Services in Cellular Networks: From GSM to 5G NR, Adrián Caldaldo García, Stefan Maier, and Abhay Phillips

MEMS-Based Integrated Navigation, Priyanka Aggarwal, Zainab Syed, Aboelmagd Noureldin, and Naser El-Sheimy

Navigation Signal Processing for GNSS Software Receivers, Thomas Pany

Navigation and Tracking in Space: Analysis and Algorithms, Sanat K. Biswas and Andrew G. Dempster

The Present and Future of Indoor Navigation, Laura Ruotsalainen, Martti Kirkko-Jaakkola, and Jukka Talvitie

Principles of GNSS, Inertial, and Multisensor Integrated Navigation Systems, Second Edition, Paul D. Groves

Radionavigation Systems, Borje Forssell

RF Positioning: Fundamentals, Applications, and Tools, Rafael Saraiva Campos and Lisandro Lovisolo

Server-Side GPS and Assisted-GPS in Java™, Neil Harper

Spread Spectrum Systems for GNSS and Wireless Communications, Jack K. Holmes

Understanding GPS/GNSS: Principles and Applications, Third Edition, Elliott Kaplan and Christopher Hegarty, editors

Ubiquitous Positioning, Robin Mannings

Wireless Positioning Technologies and Applications, Second Edition, Alan Bensky

For further information on these and other Artech House titles, including previously considered out-of-print books now available through our In-Print-Forever® (IPF®) program, contact:

Artech House Publishers
685 Canton Street
Norwood, MA 02062
Phone: 781-769-9750
Fax: 781-769-6334
e-mail: artech@artechhouse.com

Artech House Books
16 Sussex Street
London SW1V 4RW UK
Phone: +44 (0)20 7596 8750
Fax: +44 (0)20 7630 0166
e-mail:
artech-uk@artechhouse.com

Find us on the World Wide Web at: www.artechhouse.com